SRA
Connecting Math Concepts

Level E Workbook

COMPREHENSIVE EDITION

A DIRECT INSTRUCTION PROGRAM

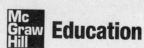
McGraw Hill Education

Bothell, WA • Chicago, IL • Columbus, OH • New York, NY

MHEonline.com

Send all inquiries to:
McGraw-Hill Education
8787 Orion Place
Columbus, OH 43240

ISBN: 978-0-02-103625-7
MHID: 0-02-103625-X

Printed in the United States of America.

5 6 7 8 9 QVS 18 17 16 15

The *McGraw·Hill* Companies

Lesson 1

Part 1

a. 17 $\xrightarrow{4}$ 21

_____

_____

_____

_____

b. N $\xrightarrow{20}$ K

_____

_____

_____

_____

Part 2

Hexagons _____

Quadrilaterals _____

Pentagons _____

Triangles _____

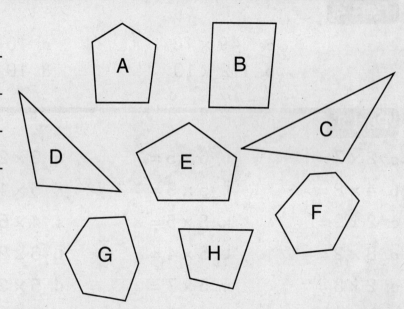

Part 3

a. $8 \times 2 =$

b. $5 \times 2 =$

c. $2 \times 10 =$

d. $2 \times 2 =$

e. $2 \times 6 =$

f. $2 \times 1 =$

g. $3 \times 2 =$

h. $2 \times 9 =$

i. $6 \times 2 =$

j. $4 \times 2 =$

k. $2 \times 3 =$

l. $7 \times 2 =$

m. $0 \times 2 =$

n. $2 \times 8 =$

o. $9 \times 2 =$

Lesson 2

Part 1

Quadrilaterals _____

Hexagons _____

Triangles _____

Pentagons _____

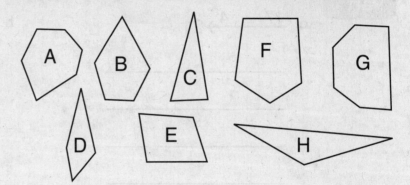

Part 2

a. 49 × 10

b. 112 × 10

c. 10 × 348

d. 10 × 400

Part 3

a. 2 × 7 =

b. 4 × 2 =

c. 2 × 9 =

d. 8 × 2 =

e. 2 × 3 =

f. 6 × 2 =

g. 2 × 0 =

h. 2 × 5 =

i. 6 × 5 =

j. 5 × 3 =

k. 8 × 5 =

l. 5 × 4 =

m. 5 × 7 =

n. 5 × 5 =

o. 5 × 1 =

p. 5 × 9 =

q. 9 × 2 =

r. 5 × 1 =

s. 4 × 5 =

t. 6 × 2 =

u. 5 × 3 =

v. 2 × 8 =

w. 5 × 6 =

x. 4 × 2 =

y. 5 × 5 =

z. 9 × 5 =

A. 0 × 2 =

B. 7 × 5 =

C. 5 × 2 =

D. 5 × 8 =

E. 3 × 2 =

F. 7 × 2 =

Part 4

a. 20 ____ 10 → 30

b. J ____ P → 6

_____

_____

_____

_____

_____

_____

Part 5

a. 5 7

b. 2 8

c. 6 4

Lesson 3

Part 1

a. $10 \times 108 =$
b. $221 \times 10 =$
c. $10 \times 506 =$
d. $10 \times 440 =$

Part 2

a. $3 \times 2 =$
b. $8 \times 5 =$
c. $2 \times 4 =$
d. $9 \times 2 =$
e. $5 \times 5 =$
f. $5 \times 9 =$
g. $2 \times 2 =$

h. $0 \times 5 =$
i. $2 \times 8 =$
j. $6 \times 2 =$
k. $7 \times 5 =$
l. $5 \times 3 =$
m. $2 \times 0 =$
n. $6 \times 5 =$

o. $10 \times 2 =$
p. $5 \times 6 =$
q. $5 \times 1 =$
r. $1 \times 2 =$
s. $2 \times 5 =$
t. $4 \times 5 =$
u. $7 \times 2 =$

Part 3

a. 34 b. 87 c. 43

Part 4

a. $\xrightarrow{\quad 5 \quad} 7$

b. $3 \xrightarrow{\quad\quad} 13$

c. $12 \xrightarrow{\quad 1 \quad} \underline{\quad}$

d. $9 \xrightarrow{\quad 3 \quad} \underline{\quad}$

e. $\xrightarrow{\quad 8 \quad} 15$

f. $5 \xrightarrow{\quad\quad} 9$

Lesson 4

Part 1

a. $560 \times 10 =$
b. $56 \times 10 =$
c. $10 \times 111 =$
d. $303 \times 10 =$
e. $10 \times 101 =$
f. $10 \times 820 =$

Part 2

a. $2 \times 6 =$
b. $8 \times 5 =$
c. $5 \times 5 =$
d. $9 \times 2 =$
e. $10 \times 5 =$
f. $2 \times 3 =$
g. $8 \times 2 =$
h. $5 \times 2 =$

i. $0 \times 2 =$
j. $1 \times 5 =$
k. $5 \times 4 =$
l. $5 \times 7 =$
m. $2 \times 2 =$
n. $2 \times 7 =$
o. $9 \times 5 =$
p. $2 \times 10 =$

q. $1 \times 2 =$
r. $5 \times 6 =$
s. $3 \times 5 =$
t. $5 \times 0 =$
u. $4 \times 2 =$
v. $2 \times 9 =$
w. $5 \times 8 =$
x. $2 \times 8 =$

Part 3

a. 35 b. 77 c. 58

Lesson 5

Part 1

a. $2 \times 8 =$
b. $7 \times 10 =$
c. $5 \times 2 =$
d. $10 \times 2 =$
e. $6 \times 5 =$
f. $10 \times 10 =$
g. $4 \times 2 =$

h. $10 \times 9 =$
i. $5 \times 10 =$
j. $7 \times 2 =$
k. $0 \times 10 =$
l. $2 \times 3 =$
m. $10 \times 6 =$
n. $5 \times 5 =$

o. $5 \times 9 =$
p. $8 \times 10 =$
q. $9 \times 2 =$
r. $5 \times 3 =$
s. $1 \times 2 =$
t. $7 \times 5 =$
u. $10 \times 2 =$

Part 2

a. $\underline{92 \quad 17}$→ ____

b. ═══ $\underline{36}$ →138

c. $\underline{49}$ ═══→179

Part 3

a. 71 c. 50

b. 27 d. 93

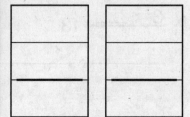

Lesson 6

Part 1

a. $7 \times 10 =$
b. $10 \times 3 =$
c. $9 \times 10 =$
d. $6 \times 10 =$
e. $4 \times 10 =$
f. $10 \times 10 =$
g. $8 \times 10 =$
h. $2 \times 10 =$
i. $10 \times 9 =$
j. $10 \times 5 =$

k. $2 \times 6 =$
l. $8 \times 5 =$
m. $7 \times 2 =$
n. $10 \times 10 =$
o. $2 \times 8 =$
p. $4 \times 2 =$
q. $0 \times 10 =$
r. $9 \times 5 =$
s. $5 \times 7 =$
t. $5 \times 5 =$

u. $3 \times 5 =$
v. $5 \times 10 =$
w. $10 \times 6 =$
x. $10 \times 5 =$
y. $4 \times 5 =$
z. $9 \times 10 =$
A. $1 \times 5 =$
B. $2 \times 9 =$
C. $10 \times 7 =$
D. $5 \times 8 =$

E. $7 \times 5 =$
F. $10 \times 1 =$
G. $5 \times 10 =$
H. $5 \times 9 =$
I. $2 \times 7 =$
J. $5 \times 6 =$
K. $5 \times 4 =$
L. $10 \times 8 =$
M. $2 \times 6 =$
N. $6 \times 5 =$

Lesson 6

Part 2

a. $\frac{3}{2}$ b. $\frac{1}{7}$ c. $\frac{9}{8}$ d. $\frac{5}{2}$ e. $\frac{7}{8}$ f. $\frac{8}{9}$

Lesson 7

Part 1

a. $3 \times 2 =$
b. $4 \times 5 =$
c. $4 \times 10 =$
d. $5 \times 2 =$
e. $5 \times 5 =$
f. $7 \times 10 =$
g. $2 \times 8 =$

h. $5 \times 9 =$
i. $10 \times 8 =$
j. $9 \times 5 =$
k. $2 \times 6 =$
l. $7 \times 5 =$
m. $10 \times 10 =$
n. $10 \times 9 =$

o. $8 \times 5 =$
p. $10 \times 5 =$
q. $5 \times 7 =$
r. $5 \times 6 =$
s. $6 \times 10 =$
t. $9 \times 2 =$
u. $10 \times 2 =$
v. $5 \times 8 =$

w. $2 \times 9 =$
x. $7 \times 2 =$
y. $6 \times 5 =$
z. $4 \times 5 =$
A. $3 \times 5 =$
B. $4 \times 2 =$
C. $2 \times 2 =$
D. $9 \times 5 =$

Part 2

a. 21   12   c. 21   24

b. 11   12   d. 112   109

Part 3

a.
```
  3 2 9
- 1 3 0
```
b.
```
  4 3 0
- 1 5 0
```
c.
```
  5 4 9
- 3 7 8
```

Part 4

a. $\underline{23 \quad Q}\rightarrow R$
R = 87

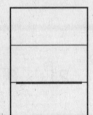

b. $\underline{47 \quad M}\rightarrow P$
M = 72

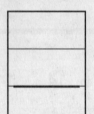

c. $\underline{19 \quad E}\rightarrow L$
L = 109

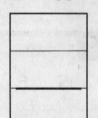

Part 5

a.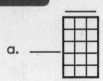

perimeter _____

b.

perimeter _____

Connecting Math Concepts

Lesson

Part 1

a. 1 × 1 =	k. 6 × 6 =	u. 2 × 6 =	E. 8 × 8 =
b. 2 × 2 =	l. 2 × 2 =	v. 3 × 3 =	F. 2 × 8 =
c. 3 × 3 =	m. 1 × 1 =	w. 4 × 4 =	G. 10 × 10 =
d. 4 × 4 =	n. 10 × 10 =	x. 2 × 10 =	H. 9 × 5 =
e. 5 × 5 =	o. 3 × 3 =	y. 5 × 2 =	I. 7 × 7 =
f. 6 × 6 =	p. 8 × 8 =	z. 5 × 7 =	J. 3 × 5 =
g. 7 × 7 =	q. 9 × 9 =	A. 7 × 5 =	K. 5 × 5 =
h. 8 × 8 =	r. 5 × 5 =	B. 9 × 2 =	L. 8 × 5 =
i. 9 × 9 =	s. 7 × 7 =	C. 9 × 9 =	M. 10 × 9 =
j. 10 × 10 =	t. 4 × 4 =	D. 5 × 6 =	N. 6 × 6 =

Part 2

a.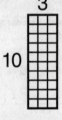

area _____

perimeter _____

b. 1 [⬚⬚⬚⬚] 4

area _____

perimeter _____

c. 3 [⬚⬚⬚⬚⬚] 5

area _____

perimeter _____

Part 3

a. 5   6   b. 14   14   c. 21   8

Part 4

a. $\frac{4}{5}$ 1   b. $\frac{7}{9}$ 1   d. $\frac{4}{1}$ 1   f. $\frac{3}{2}$ 1   h. $\frac{5}{5}$ 1

c. $\frac{10}{10}$ 1   e. $\frac{1}{2}$ 1   g. $\frac{5}{6}$ 1   i. $\frac{7}{4}$ 1

Copyright © The McGraw-Hill Companies, Inc.

Connecting Math Concepts

Lesson 9

Part 1

a. $7 \times 7 =$
b. $2 \times 2 =$
c. $5 \times 5 =$
d. $8 \times 8 =$
e. $3 \times 3 =$
f. $9 \times 9 =$
g. $6 \times 6 =$
h. $4 \times 4 =$
i. $10 \times 10 =$
j. $1 \times 1 =$

k. $10 \times 9 =$
l. $7 \times 10 =$
m. $4 \times 4 =$
n. $2 \times 4 =$
o. $7 \times 7 =$
p. $6 \times 5 =$
q. $8 \times 2 =$
r. $6 \times 6 =$
s. $5 \times 2 =$
t. $5 \times 10 =$

u. $10 \times 10 =$
v. $3 \times 3 =$
w. $6 \times 2 =$
x. $5 \times 9 =$
y. $8 \times 8 =$
z. $7 \times 5 =$
A. $5 \times 8 =$
B. $2 \times 9 =$
C. $9 \times 9 =$
D. $2 \times 10 =$

E. $5 \times 6 =$
F. $2 \times 2 =$
G. $9 \times 5 =$
H. $5 \times 7 =$
I. $10 \times 6 =$
J. $5 \times 5 =$
K. $2 \times 7 =$
L. $8 \times 5 =$
M. $8 \times 10 =$
N. $5 \times 4 =$

Part 2

a. $\frac{5}{3}$ 1

b. $\frac{14}{14}$ 1

c. $\frac{7}{9}$ 1

d. $\frac{7}{1}$ 1

e. $\frac{1}{2}$ 1

f. $\frac{6}{5}$ 1

g. $\frac{3}{3}$ 1

Part 3

a.

area _____

perimeter _____

b.

area _____

perimeter _____

Part 4

a. $16 \xrightarrow{4} 64$

b. $N \xrightarrow{20} K$

Lesson 10

Part 1

a.	7 × 7 =	k.	5 × 4 =	u.	10 × 8 =	E.	3 × 5 =
b.	4 × 4 =	l.	7 × 5 =	v.	7 × 2 =	F.	6 × 5 =
c.	1 × 1 =	m.	5 × 8 =	w.	10 × 9 =	G.	8 × 8 =
d.	8 × 8 =	n.	10 × 2 =	x.	5 × 9 =	H.	8 × 2 =
e.	5 × 5 =	o.	10 × 10 =	y.	2 × 9 =	I.	5 × 6 =
f.	2 × 2 =	p.	9 × 5 =	z.	2 × 4 =	J.	7 × 7 =
g.	9 × 9 =	q.	9 × 2 =	A.	3 × 2 =	K.	9 × 10 =
h.	6 × 6 =	r.	2 × 6 =	B.	4 × 4 =	L.	6 × 6 =
i.	3 × 3 =	s.	5 × 10 =	C.	5 × 7 =	M.	8 × 10 =
j.	10 × 10 =	t.	8 × 5 =	D.	9 × 9 =	N.	5 × 5 =

Part 2

a. 1   c. 1   e. 1

b. 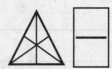 1   d. 1   f. 1

Part 3

a.    b.   c.    d.

Independent Work

Part 4

a.   b. 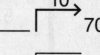   c.

area _____   area _____   area _____

perimeter _____   perimeter _____   perimeter _____

Lesson 11

Part 1

a. 9 × 1 =	k. 9 × 2 =	u. 5 × 5 =	E. 4 × 9 =
b. 9 × 2 =	l. 9 × 7 =	v. 8 × 9 =	F. 8 × 8 =
c. 9 × 3 =	m. 9 × 5 =	w. 7 × 2 =	G. 8 × 10 =
d. 9 × 4 =	n. 9 × 8 =	x. 9 × 2 =	H. 4 × 4 =
e. 9 × 5 =	o. 9 × 9 =	y. 8 × 5 =	I. 4 × 2 =
f. 9 × 6 =	p. 9 × 4 =	z. 9 × 5 =	J. 7 × 7 =
g. 9 × 7 =	q. 9 × 3 =	A. 3 × 5 =	K. 3 × 9 =
h. 9 × 8 =	r. 9 × 6 =	B. 2 × 8 =	L. 7 × 9 =
i. 9 × 9 =	s. 9 × 10 =	C. 9 × 9 =	M. 10 × 10 =
j. 9 × 10 =	t. 9 × 1 =	D. 5 × 10 =	N. 6 × 9 =

Part 2

a. 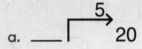   b.   c.   d.

Part 3

a.  1   d. 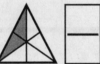 1

b. 1   e. 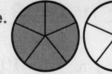 1

c. 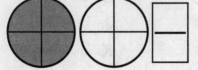 1   f. 1

Lesson 12

Part 1

a. 9 × 1 =	k. 8 × 9 =	u. 8 × 5 =	E. 7 × 8 =
b. 9 × 2 =	l. 10 × 9 =	v. 9 × 8 =	F. 4 × 9 =
c. 9 × 3 =	m. 9 × 7 =	w. 5 × 5 =	G. 7 × 5 =
d. 9 × 4 =	n. 9 × 4 =	x. 9 × 9 =	H. 9 × 5 =
e. 9 × 5 =	o. 9 × 2 =	y. 2 × 8 =	I. 7 × 9 =
f. 9 × 6 =	p. 1 × 9 =	z. 5 × 9 =	J. 2 × 7 =
g. 9 × 7 =	q. 9 × 9 =	A. 5 × 7 =	K. 6 × 9 =
h. 9 × 8 =	r. 5 × 9 =	B. 8 × 8 =	L. 10 × 10 =
i. 9 × 9 =	s. 9 × 3 =	C. 5 × 6 =	M. 4 × 5 =
j. 9 × 10 =	t. 6 × 9 =	D. 4 × 4 =	N. 6 × 6 =

Part 2

a.  1   c. 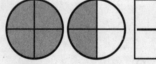 1   e. 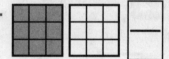 1

b. 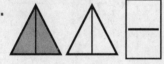 1   d. 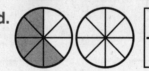 1   f.  1

Lesson 13

Part 1

a. 9 × 1 =	k. 9 × 7 =	u. 4 × 9 =	E. 9 × 2 =
b. 5 × 9 =	l. 2 × 2 =	v. 6 × 6 =	F. 5 × 9 =
c. 10 × 9 =	m. 9 × 9 =	w. 9 × 5 =	G. 3 × 3 =
d. 3 × 9 =	n. 10 × 10 =	x. 9 × 3 =	H. 9 × 8 =
e. 9 × 8 =	o. 9 × 6 =	y. 8 × 5 =	I. 6 × 9 =
f. 9 × 9 =	p. 9 × 5 =	z. 5 × 5 =	J. 0 × 9 =
g. 9 × 4 =	q. 10 × 9 =	A. 8 × 9 =	K. 3 × 9 =
h. 7 × 9 =	r. 1 × 9 =	B. 8 × 8 =	L. 4 × 4 =
i. 9 × 6 =	s. 5 × 7 =	C. 7 × 7 =	M. 9 × 4 =
j. 2 × 9 =	t. 2 × 9 =	D. 7 × 9 =	N. 9 × 10 =

Lesson 13

Part 2

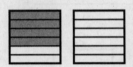 a.

c. 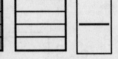 1

b. 1   d. 1

Part 3

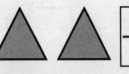

3	9
12	18
24	27
30	

Lesson 14

Part 1

a. $3 \times 3 =$ g. $2 \times 3 =$ m. $6 \times 3 =$ s. $9 \times 5 =$

b. $3 \times 1 =$ h. $3 \times 4 =$ n. $7 \times 9 =$ t. $9 \times 9 =$

c. $3 \times 5 =$ i. $1 \times 3 =$ o. $8 \times 8 =$ u. $3 \times 6 =$

d. $3 \times 2 =$ j. $6 \times 3 =$ p. $5 \times 5 =$ v. $8 \times 9 =$

e. $3 \times 4 =$ k. $3 \times 5 =$ q. $10 \times 10 =$ w. $7 \times 7 =$

f. $3 \times 6 =$ l. $3 \times 3 =$ r. $4 \times 6 =$ x. $4 \times 9 =$

Part 2

c. $\frac{8}{8}$ 1   f. $\frac{2}{1}$ 1   i. $\frac{6}{5}$ 1

a. $\frac{5}{7}$ 1   d. $\frac{10}{6}$ 1   g. $\frac{3}{3}$ 1   j. $\frac{7}{12}$ 1

b. $\frac{9}{4}$ 1   e. $\frac{12}{15}$ 1   h. $\frac{17}{17}$ 1   k. $\frac{7}{2}$ 1

Lesson 14

a. 9 ⌐→ 45

d. ___ ⌐→² 18

g. ___ ⌐→¹ 9

_____

_____

_____

b. ___ ⌐→³ 27

e. 9 ⌐→ 36

h. ___ ⌐→² 12

_____

_____

_____

c. 5 ⌐→⁴ ___

f. 9 ⌐→³ ___

i. 9 ⌐→⁴ ___

_____

_____

_____

Lesson 15

Part 1

a. $3 \times 9 =$
b. $3 \times 6 =$
c. $3 \times 8 =$
d. $9 \times 3 =$
e. $10 \times 3 =$
f. $7 \times 3 =$
g. $3 \times 5 =$
h. $8 \times 3 =$

i. $3 \times 7 =$
j. $6 \times 3 =$
k. $10 \times 3 =$
l. $7 \times 3 =$
m. $3 \times 5 =$
n. $0 \times 3 =$
o. $3 \times 4 =$
p. $1 \times 3 =$

q. $3 \times 2 =$
r. $6 \times 3 =$
s. $3 \times 8 =$
t. $3 \times 1 =$
u. $2 \times 3 =$
v. $3 \times 6 =$
w. $4 \times 3 =$
x. $3 \times 10 =$

y. $9 \times 3 =$
z. $3 \times 7 =$
A. $3 \times 3 =$
B. $5 \times 3 =$
C. $8 \times 3 =$
D. $3 \times 9 =$

Connecting Math Concepts

Lesson 15

a. 3 ⟶³ ___

d. 5 ⟹ 20

g. ___ ⟶⁷ 35

j. ___ ⟶⁴ 36

b. ___ ⟶⁹ 9

e. 9 ⟶⁴ ___

h. ___ ⟶² 18

k. 9 ⟶⁵ ___

c. ___ ⟶² 14

f. ___ ⟶³ 27

i. 6 ⟹ 12

l. 3 ⟹ 9

Lesson 16

Part 1

a. 3 ⟹ 9

d. ___ ⟶⁹ 36

g. ___ ⟶² 14

b. 9 ⟶³ ___

e. ___ ⟶² 18

h. ___ ⟶³ 27

c. 9 ⟹ 9

f. 9 ⟶⁵ ___

i. ___ ⟶⁴ 36

Lesson 16

Part 2

a. $4\overline{)36}$	g. $2\overline{)20}$	m. $9\overline{)9}$	s. $6\overline{)12}$	y. $3\overline{)9}$
b. $5\overline{)35}$	h. $9\overline{)45}$	n. $9\overline{)45}$	t. $7\overline{)35}$	z. $4\overline{)20}$
c. $5\overline{)45}$	i. $3\overline{)9}$	o. $4\overline{)8}$	u. $5\overline{)45}$	A. $1\overline{)9}$
d. $9\overline{)18}$	j. $2\overline{)14}$	p. $10\overline{)80}$	v. $2\overline{)6}$	B. $8\overline{)16}$
e. $3\overline{)27}$	k. $9\overline{)36}$	q. $2\overline{)16}$	w. $2\overline{)12}$	C. $2\overline{)10}$
f. $5\overline{)25}$	l. $2\overline{)18}$	r. $9\overline{)27}$	x. $2\overline{)10}$	D. $3\overline{)6}$

Part 3

a. $3 \times 3 =$	k. $2 \times 3 =$	u. $4 \times 9 =$	E. $4 \times 10 =$
b. $4 \times 3 =$	l. $5 \times 5 =$	v. $10 \times 10 =$	F. $0 \times 7 =$
c. $3 \times 9 =$	m. $6 \times 6 =$	w. $8 \times 1 =$	G. $7 \times 2 =$
d. $1 \times 3 =$	n. $7 \times 3 =$	x. $9 \times 2 =$	H. $2 \times 9 =$
e. $9 \times 4 =$	o. $9 \times 5 =$	y. $9 \times 3 =$	I. $4 \times 5 =$
f. $2 \times 9 =$	p. $3 \times 3 =$	z. $4 \times 4 =$	J. $9 \times 3 =$
g. $5 \times 3 =$	q. $3 \times 6 =$	A. $7 \times 5 =$	K. $3 \times 4 =$
h. $2 \times 3 =$	r. $1 \times 9 =$	B. $8 \times 2 =$	L. $5 \times 1 =$
i. $3 \times 0 =$	s. $6 \times 5 =$	C. $3 \times 5 =$	M. $2 \times 4 =$
j. $5 \times 4 =$	t. $5 \times 8 =$	D. $2 \times 6 =$	N. $1 \times 7 =$

Lesson 17

Part 1

a. $9 \Rightarrow 54$

b. $\underline{\quad} \xrightarrow{3} 27$

c. $3 \xrightarrow{3} \underline{\quad}$

d. $\underline{\quad} \xrightarrow{5} 10$

e. $9 \xrightarrow{6} \underline{\quad}$

f. $\underline{\quad} \xrightarrow{2} 16$

g. $\underline{\quad} \xrightarrow{4} 36$

h. $\underline{\quad} \xrightarrow{3} 15$

i. $\underline{\quad} \xrightarrow{1} 9$

j. $9 \Rightarrow 27$

k. $9 \xrightarrow{4} \underline{\quad}$

l. $\underline{\quad} \xrightarrow{6} 54$

Part 2

a. $6 \overline{)54}$ e. $5 \overline{)45}$ i. $3 \overline{)9}$ m. $10 \overline{)70}$ q. $8 \overline{)16}$ u. $10 \overline{)10}$

b. $9 \overline{)36}$ f. $2 \overline{)16}$ j. $9 \overline{)54}$ n. $9 \overline{)45}$ r. $9 \overline{)27}$ v. $2 \overline{)6}$

c. $2 \overline{)8}$ g. $3 \overline{)27}$ k. $4 \overline{)36}$ o. $4 \overline{)8}$ s. $6 \overline{)12}$ w. $5 \overline{)25}$

d. $9 \overline{)18}$ h. $1 \overline{)9}$ l. $2 \overline{)10}$ p. $5 \overline{)20}$ t. $3 \overline{)15}$ x. $2 \overline{)18}$

Part 3

a. $9 \Rightarrow 18$

b. $9 \Rightarrow 18$

c. $9 \xrightarrow{4} \underline{\quad}$

d. $9 \xrightarrow{4} \underline{\quad}$

e. $\underline{\quad} \xrightarrow{2} 8$

f. $\underline{\quad} \xrightarrow{2} 8$

g. $5 \Rightarrow 15$

h. $5 \Rightarrow 15$

i. $\underline{\quad} \xrightarrow{1} 9$

j. $\underline{\quad} \xrightarrow{1} 9$

Lesson 17

Part 4

a. 3 × 1 =	k. 3 × 5 =	u. 4 × 4 =	E. 9 × 5 =
b. 3 × 2 =	l. 3 × 2 =	v. 3 × 7 =	F. 5 × 5 =
c. 3 × 3 =	m. 3 × 9 =	w. 9 × 2 =	G. 9 × 6 =
d. 3 × 4 =	n. 3 × 6 =	x. 8 × 5 =	H. 3 × 9 =
e. 3 × 5 =	o. 3 × 4 =	y. 2 × 2 =	I. 9 × 1 =
f. 3 × 6 =	p. 3 × 7 =	z. 3 × 6 =	J. 10 × 10 =
g. 3 × 7 =	q. 3 × 8 =	A. 5 × 6 =	K. 3 × 5 =
h. 3 × 8 =	r. 3 × 1 =	B. 7 × 5 =	L. 9 × 3 =
i. 3 × 9 =	s. 3 × 10 =	C. 4 × 9 =	M. 4 × 3 =
j. 3 × 10 =	t. 3 × 3 =	D. 6 × 9 =	N. 9 × 4 =

Lesson 18

Part 1

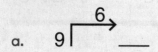

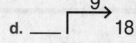

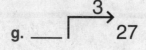

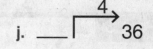

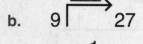

Part 2

a. 4⟌36	e. 9⟌27	i. 1⟌1	m. 1⟌8	q. 2⟌16	u. 5⟌45
b. 1⟌5	f. 6⟌54	j. 7⟌14	n. 3⟌27	r. 3⟌9	v. 2⟌6
c. 9⟌45	g. 10⟌40	k. 9⟌36	o. 9⟌54	s. 9⟌36	w. 2⟌18
d. 9⟌18	h. 3⟌15	l. 5⟌40	p. 7⟌70	t. 5⟌20	x. 6⟌30

Connecting Math Concepts

Lesson 18

Part 3

a. $2 \xrightarrow{\quad 6\quad}$ ___

d. $\underequals{\quad 2\quad} 6$

g. $\underbracket{\quad}\xrightarrow{\;3\;} 27$

j. $9 \xrightarrow{\quad 2\quad}$ ___

___

___

___

___

b. $2\underbracket{\;\xrightarrow{\;6\;}\;}$ ___

e. $9\underbracket{\;\xrightarrow{\;5\;}\;}$ ___

h. $5\underbracket{\;\xrightarrow{\;}\;} 20$

k. $9\underbracket{\;\xrightarrow{\;2\;}\;}$ ___

___

___

___

___

c. $2\underbracket{\;\xrightarrow{\;}\;} 6$

f. $9 \xrightarrow{\quad 5\quad}$ ___

i. $\underequals{\quad 10\quad} 20$

l. $9\underbracket{\;\xrightarrow{\;}\;} 36$

___

___

___

___

Part 4

a. $3 = \dfrac{}{10} = \dfrac{}{5} = \dfrac{}{9} = \dfrac{}{2}$

b. $2 = \dfrac{}{6} = \dfrac{}{4} = \dfrac{}{8} = \dfrac{}{1}$

Part 5

a. $6 \times 9 =$
b. $9 \times 10 =$
c. $2 \times 9 =$
d. $3 \times 3 =$
e. $4 \times 9 =$
f. $4 \times 5 =$
g. $2 \times 1 =$
h. $3 \times 9 =$

i. $10 \times 2 =$
j. $5 \times 9 =$
k. $6 \times 5 =$
l. $0 \times 8 =$
m. $2 \times 2 =$
n. $10 \times 6 =$
o. $8 \times 5 =$
p. $6 \times 2 =$

q. $8 \times 2 =$
r. $5 \times 5 =$
s. $3 \times 10 =$
t. $9 \times 6 =$
u. $5 \times 7 =$
v. $6 \times 1 =$
w. $9 \times 4 =$
x. $5 \times 10 =$

y. $9 \times 3 =$
z. $1 \times 1 =$
A. $5 \times 3 =$
B. $9 \times 5 =$
C. $10 \times 4 =$
D. $2 \times 5 =$
E. $1 \times 9 =$
F. $8 \times 0 =$

Lesson 19

Part 1

a. 9 ⌐5→ _____   d. _____ ⌐3→ 9   g. _____ ⌐3→ 27   j. 9 ⌐2→ _____

b. 9 ⌐=→ 54   e. 9 ⌐=→ 36   h. 6 ⌐=→ 12   k. _____ ⌐6→ 54

c. _____ ⌐9→ 9   f. 9 ⌐6→ _____   i. _____ ⌐4→ 36   l. 9 ⌐=→ 27

Part 2

a. 5⟌15   e. 1⟌9   i. 8⟌0   m. 7⟌7   q. 9⟌27   u. 4⟌36

b. 6⟌54   f. 3⟌9   j. 9⟌18   n. 9⟌54   r. 5⟌25   v. 10⟌30

c. 6⟌12   g. 3⟌27   k. 2⟌8   o. 5⟌30   s. 2⟌4   w. 8⟌40

d. 9⟌36   h. 5⟌35   l. 9⟌45   p. 6⟌60   t. 4⟌20   x. 2⟌18

Part 3

30	20	
39		79

a.

b.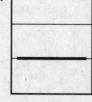

c. 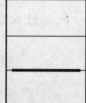   d.    e.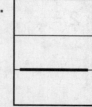

Connecting Math Concepts

Lesson 19

Part 4

a. $3 \times 3 =$ i. $10 \times 10 =$ q. $5 \times 1 =$ y. $4 \times 5 =$

b. $9 \times 10 =$ j. $3 \times 3 =$ r. $5 \times 9 =$ z. $9 \times 3 =$

c. $5 \times 3 =$ k. $2 \times 6 =$ s. $8 \times 2 =$ A. $1 \times 1 =$

d. $4 \times 9 =$ l. $0 \times 4 =$ t. $3 \times 2 =$ B. $7 \times 5 =$

e. $9 \times 3 =$ m. $3 \times 9 =$ u. $1 \times 7 =$ C. $9 \times 2 =$

f. $6 \times 5 =$ n. $9 \times 6 =$ v. $3 \times 5 =$ D. $2 \times 5 =$

g. $6 \times 9 =$ o. $5 \times 5 =$ w. $2 \times 9 =$ E. $1 \times 8 =$

h. $9 \times 2 =$ p. $3 \times 10 =$ x. $3 \times 0 =$ F. $2 \times 4 =$

Lesson 20

Part 1

a. $6 \xrightarrow{3}$ ___   d. $4 \xrightarrow{} 12$ g. $9 \xrightarrow{6}$ ___   j. $3 \xrightarrow{3}$ ___

b. ___ $\xrightarrow{6} 54$ e. ___ $\xrightarrow{3} 9$ h. $6 \xrightarrow{} 18$ k. ___ $\xrightarrow{4} 36$

c. ___ $\xrightarrow{5} 15$ f. ___ $\xrightarrow{3} 18$ i. ___ $\xrightarrow{3} 12$ l. $4 \xrightarrow{3}$ ___

Part 2

a. $4\overline{)12}$ f. $3\overline{)18}$ k. $6\overline{)30}$ p. $10\overline{)30}$ u. $5\overline{)15}$ z. $2\overline{)14}$

b. $3\overline{)27}$ g. $3\overline{)0}$ l. $3\overline{)12}$ q. $7\overline{)14}$ v. $1\overline{)9}$ A. $6\overline{)6}$

c. $6\overline{)54}$ h. $9\overline{)45}$ m. $4\overline{)36}$ r. $9\overline{)27}$ w. $5\overline{)25}$ B. $9\overline{)54}$

d. $5\overline{)40}$ i. $6\overline{)18}$ n. $2\overline{)10}$ s. $4\overline{)8}$ x. $2\overline{)6}$ C. $5\overline{)50}$

e. $9\overline{)36}$ j. $2\overline{)18}$ o. $5\overline{)20}$ t. $7\overline{)35}$ y. $3\overline{)9}$ D. $1\overline{)4}$

Connecting Math Concepts

Lesson 20

Part 3

	12	
13		32
29		

a.

b.

c.   d.   e.

Part 4

a. 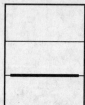 1 → 9

d. 9 → 4 → __

g. 10 → 2 → __

j. __ ⌐2→ 10

b. __ ⌐1→ 9

e. 9 → → 18

h. 10 ⌐2→ __

k. 9 ⌐→ 27

c. 9 ⌐4→ __

f. 9 ⌐→ 18

i. 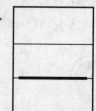 2 → 10

l. 2 → 4

Part 5

a. 4 × 9 =	g. 3 × 3 =	m. 9 × 10 =	s. 9 × 3 =
b. 6 × 3 =	h. 8 × 0 =	n. 6 × 9 =	t. 6 × 5 =
c. 3 × 4 =	i. 3 × 9 =	o. 5 × 3 =	u. 10 × 6 =
d. 5 × 8 =	j. 3 × 6 =	p. 2 × 2 =	v. 9 × 4 =
e. 8 × 2 =	k. 2 × 7 =	q. 0 × 9 =	w. 6 × 2 =
f. 9 × 6 =	l. 4 × 3 =	r. 1 × 6 =	x. 9 × 1 =

Connecting Math Concepts

Lesson 21

Part 1

a. 4 ⟶ 12

b. 3 ⟶³ ___

c. 9 ⟶ 36

d. ___ ⟶⁶ 54

e. 9 ⟶ 27

f. ___ ⟶² 18

g. 3 ⟶ 18

h. 3 ⟶ 9

i. 3 ⟶⁶ ___

j. ___ ⟶⁴ 12

k. 9 ⟶⁴ ___

l. ___ ⟶³ 27

Part 2

a. 3$\overline{)12}$ f. 3$\overline{)9}$ k. 9$\overline{)36}$ p. 10$\overline{)30}$ u. 5$\overline{)15}$ z. 10$\overline{)100}$

b. 9$\overline{)27}$ g. 9$\overline{)9}$ l. 9$\overline{)18}$ q. 7$\overline{)14}$ v. 2$\overline{)8}$ A. 6$\overline{)6}$

c. 6$\overline{)54}$ h. 4$\overline{)12}$ m. 9$\overline{)36}$ r. 9$\overline{)18}$ w. 5$\overline{)25}$ B. 9$\overline{)54}$

d. 3$\overline{)18}$ i. 4$\overline{)0}$ n. 3$\overline{)30}$ s. 1$\overline{)4}$ x. 5$\overline{)10}$ C. 2$\overline{)4}$

e. 5$\overline{)30}$ j. 6$\overline{)18}$ o. 4$\overline{)36}$ t. 10$\overline{)60}$ y. 9$\overline{)0}$ D. 3$\overline{)27}$

Part 3

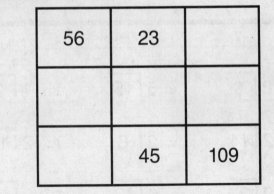

56	23	
	45	109

a.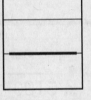

b.

c. 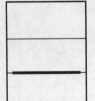 d. e.

Part 4

a. 3$\overline{)693}$

b. 2$\overline{)684}$

c. 4$\overline{)804}$

d. 2$\overline{)480}$

Lesson 21

Part 5

a. $4 \times 3 =$ i. $9 \times 2 =$ q. $5 \times 7 =$ y. $2 \times 9 =$

b. $9 \times 6 =$ j. $0 \times 6 =$ r. $6 \times 9 =$ z. $1 \times 6 =$

c. $8 \times 5 =$ k. $10 \times 8 =$ s. $4 \times 5 =$ A. $6 \times 5 =$

d. $3 \times 6 =$ l. $5 \times 9 =$ t. $3 \times 1 =$ B. $9 \times 4 =$

e. $4 \times 9 =$ m. $6 \times 2 =$ u. $2 \times 8 =$ C. $10 \times 10 =$

f. $7 \times 2 =$ n. $3 \times 4 =$ v. $6 \times 3 =$ D. $3 \times 9 =$

g. $9 \times 3 =$ o. $1 \times 8 =$ w. $5 \times 9 =$ E. $9 \times 0 =$

h. $3 \times 3 =$ p. $2 \times 4 =$ x. $4 \times 10 =$ F. $3 \times 2 =$

Lesson 22

Part 1

a. $6 \longrightarrow 18$ d. $\_\_ \xrightarrow{6} 12$ g. $\_\_ \xrightarrow{3} 9$ j. $9 \longrightarrow 54$

b. $\_\_ \xrightarrow{6} 54$ e. $\_\_ \xrightarrow{3} 12$ h. $9 \longrightarrow 36$ k. $\_\_ \xrightarrow{3} 18$

c. $6 \longrightarrow 0$ f. $9 \xrightarrow{3} \_\_$ i. $4 \xrightarrow{3} \_\_$ l. $9 \longrightarrow 45$

Part 2

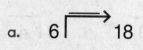

a. $9\overline{)54}$ f. $2\overline{)0}$ k. $5\overline{)30}$ p. $8\overline{)0}$ u. $5\overline{)15}$ z. $1\overline{)2}$

b. $3\overline{)12}$ g. $4\overline{)20}$ l. $3\overline{)27}$ q. $2\overline{)14}$ v. $2\overline{)8}$ A. $2\overline{)18}$

c. $3\overline{)18}$ h. $9\overline{)45}$ m. $2\overline{)4}$ r. $9\overline{)27}$ w. $6\overline{)12}$ B. $6\overline{)54}$

d. $4\overline{)36}$ i. $2\overline{)12}$ n. $8\overline{)40}$ s. $3\overline{)27}$ x. $6\overline{)30}$ C. $5\overline{)10}$

e. $10\overline{)10}$ j. $4\overline{)12}$ o. $9\overline{)36}$ t. $5\overline{)35}$ y. $3\overline{)9}$ D. $10\overline{)100}$

Lesson 22

Part 3

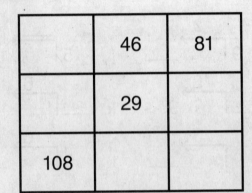

a.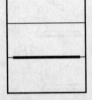

b.

c.

d.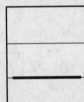

e.

Part 4

a. $3\overline{)279}$

b. $2\overline{)120}$

c. $5\overline{)355}$

Part 5

a. $0 \times 20 =$
b. $4 \times 3 =$
c. $10 \times 9 =$
d. $3 \times 6 =$
e. $1 \times 8 =$
f. $9 \times 3 =$
g. $7 \times 5 =$
h. $4 \times 9 =$

i. $3 \times 5 =$
j. $9 \times 6 =$
k. $10 \times 4 =$
l. $3 \times 3 =$
m. $6 \times 3 =$
n. $4 \times 2 =$
o. $9 \times 5 =$
p. $6 \times 2 =$

q. $9 \times 4 =$
r. $5 \times 8 =$
s. $6 \times 10 =$
t. $2 \times 9 =$
u. $3 \times 4 =$
v. $6 \times 5 =$
w. $1 \times 10 =$
x. $6 \times 9 =$

y. $3 \times 9 =$
z. $6 \times 0 =$
A. $2 \times 10 =$
B. $5 \times 5 =$
C. $5 \times 4 =$
D. $7 \times 2 =$
E. $4 \times 1 =$
F. $2 \times 3 =$

Part 6

a.
$$\begin{array}{r} \overset{3}{1}7 \\ \times\ 5 \\ \hline 5 \end{array}$$

b.
$$\begin{array}{r} \overset{4}{2}5 \\ \times\ 9 \\ \hline 5 \end{array}$$

c.
$$\begin{array}{r} \overset{4}{3}8 \\ \times\ 5 \\ \hline 0 \end{array}$$

d.
$$\begin{array}{r} \overset{1}{5}2 \\ \times\ 6 \\ \hline 2 \end{array}$$

Lesson 23

Part 1

a. $6 \overset{3}{\longrightarrow} \underline{\quad}$ d. $9 \overset{=}{\longrightarrow} 18$ g. $9 \overset{4}{\longrightarrow} \underline{\quad}$ j. $5 \overset{=}{\longrightarrow} 15$

b. $9 \overset{=}{\longrightarrow} 36$ e. $6 \overset{=}{\longrightarrow} 18$ h. $\underline{\quad} \overset{3}{\longrightarrow} 9$ k. $\underline{\quad} \overset{6}{\longrightarrow} 12$

c. $\underline{\quad} \overset{3}{\longrightarrow} 12$ f. $\underline{\quad} \overset{6}{\longrightarrow} 54$ i. $9 \overset{3}{\longrightarrow} \underline{\quad}$ l. $9 \overset{6}{\longrightarrow} \underline{\quad}$

Part 2

a. $1\overline{)4}$ f. $2\overline{)18}$ k. $9\overline{)36}$ p. $3\overline{)15}$ u. $7\overline{)70}$ z. $6\overline{)18}$

b. $6\overline{)54}$ g. $3\overline{)18}$ l. $6\overline{)0}$ q. $8\overline{)16}$ v. $11\overline{)0}$ A. $2\overline{)8}$

c. $3\overline{)9}$ h. $6\overline{)30}$ m. $6\overline{)12}$ r. $1\overline{)9}$ w. $5\overline{)40}$ B. $9\overline{)45}$

d. $2\overline{)12}$ i. $9\overline{)27}$ n. $3\overline{)12}$ s. $9\overline{)54}$ x. $3\overline{)27}$ C. $10\overline{)30}$

e. $4\overline{)12}$ j. $6\overline{)6}$ o. $4\overline{)36}$ t. $3\overline{)6}$ y. $9\overline{)18}$ D. $5\overline{)25}$

Part 3

a.
$$\begin{array}{r} \overset{3}{2}9 \\ \times\ 4 \\ \hline 6 \end{array}$$

b.
$$\begin{array}{r} \overset{4}{4}5 \\ \times\ 9 \\ \hline 5 \end{array}$$

c.
$$\begin{array}{r} 59 \\ \times\ 3 \\ \hline \end{array}$$

d.
$$\begin{array}{r} 25 \\ \times\ 8 \\ \hline \end{array}$$

Part 4

a. $2\overline{)166}$ c. $3\overline{)276}$ e. $3\overline{)690}$

b. $2\overline{)806}$ d. $9\overline{)180}$ f. $2\overline{)148}$

Connecting Math Concepts

Lesson 23

Part 5

a. $3 \times 6 =$ g. $10 \times 0 =$ m. $3 \times 4 =$ s. $2 \times 2 =$ y. $6 \times 2 =$

b. $9 \times 6 =$ h. $9 \times 3 =$ n. $8 \times 5 =$ t. $5 \times 9 =$ z. $5 \times 6 =$

c. $4 \times 3 =$ i. $5 \times 3 =$ o. $9 \times 4 =$ u. $6 \times 3 =$ A. $1 \times 5 =$

d. $4 \times 9 =$ j. $9 \times 2 =$ p. $1 \times 9 =$ v. $1 \times 1 =$ B. $9 \times 6 =$

e. $3 \times 3 =$ k. $7 \times 1 =$ q. $5 \times 10 =$ w. $10 \times 3 =$ C. $0 \times 4 =$

f. $6 \times 9 =$ l. $10 \times 8 =$ r. $2 \times 8 =$ x. $5 \times 7 =$ D. $5 \times 4 =$

Lesson 24

Part 1

a. $9 \rightarrow 54$ d. $9 \rightarrow 36$ g. $3 \xrightarrow{3} \underline{\ \ }$ j. $\underline{\ \ } \xrightarrow{9} 9$

b. $6 \xrightarrow{3} \underline{\ \ }$ e. $\underline{\ \ } \xrightarrow{3} 12$ h. $9 \rightarrow 27$ k. $4 \xrightarrow{3} \underline{\ \ }$

c. $\underline{\ \ } \xrightarrow{3} 6$ f. $9 \rightarrow 18$ i. $\underline{\ \ } \xrightarrow{3} 15$ l. $\underline{\ \ } \xrightarrow{3} 18$

Part 2

a. $4\overline{)12}$ f. $6\overline{)54}$ k. $4\overline{)40}$ p. $3\overline{)30}$ u. $10\overline{)20}$ z. $9\overline{)45}$

b. $6\overline{)12}$ g. $3\overline{)18}$ l. $9\overline{)27}$ q. $8\overline{)16}$ v. $3\overline{)15}$ A. $10\overline{)10}$

c. $1\overline{)9}$ h. $9\overline{)18}$ m. $2\overline{)12}$ r. $6\overline{)18}$ w. $2\overline{)14}$ B. $2\overline{)10}$

d. $3\overline{)9}$ i. $4\overline{)36}$ n. $3\overline{)12}$ s. $2\overline{)18}$ x. $1\overline{)6}$ C. $3\overline{)0}$

e. $9\overline{)0}$ j. $5\overline{)40}$ o. $6\overline{)30}$ t. $5\overline{)20}$ y. $2\overline{)6}$ D. $4\overline{)8}$

Part 3

a.
$$\begin{array}{r} 29 \\ \times\ 3 \\ \hline \end{array}$$

b.
$$\begin{array}{r} 15 \\ \times\ 8 \\ \hline \end{array}$$

c.
$$\begin{array}{r} 74 \\ \times\ 5 \\ \hline \end{array}$$

d.
$$\begin{array}{r} 29 \\ \times\ 4 \\ \hline \end{array}$$

Lesson 24

Part 4

a. 6 × 9 =
b. 6 × 2 =
c. 6 × 5 =
d. 6 × 3 =
e. 4 × 9 =
f. 4 × 3 =

g. 4 × 5 =
h. 4 × 2 =
i. 9 × 3 =
j. 9 × 0 =
k. 9 × 5 =
l. 9 × 2 =

m. 9 × 1 =
n. 3 × 2 =
o. 3 × 5 =
p. 3 × 3 =
q. 3 × 9 =
r. 3 × 0 =

s. 3 × 10 =
t. 3 × 4 =
u. 1 × 2 =
v. 5 × 2 =
w. 5 × 5 =
x. 5 × 7 =

y. 2 × 7 =
z. 10 × 7 =
A. 10 × 1 =
B. 7 × 1 =
C. 9 × 6 =
D. 9 × 4 =

Lesson 25

Part 1

a. 4 × 3 =
b. 5 × 6 =
c. 2 × 9 =
d. 6 × 9 =
e. 3 × 6 =
f. 0 × 10 =

g. 9 × 3 =
h. 10 × 10 =
i. 4 × 1 =
j. 7 × 5 =
k. 3 × 3 =
l. 9 × 4 =

m. 3 × 10 =
n. 1 × 5 =
o. 2 × 4 =
p. 6 × 3 =
q. 5 × 8 =
r. 4 × 0 =

s. 6 × 2 =
t. 3 × 4 =
u. 5 × 9 =
v. 1 × 1 =
w. 4 × 10 =
x. 5 × 5 =

y. 10 × 6 =
z. 4 × 5 =
A. 2 × 3 =
B. 9 × 6 =
C. 4 × 2 =
D. 4 × 9 =

Part 2

a.
```
  270
 -182
```

b.
```
  613
 - 54
```

c.
```
  582
 -394
```

d.
```
  461
 - 94
```

Part 3

a. 3)27
b. 4)12
c. 2)18
d. 6)18
e. 2)12

f. 10)0
g. 5)10
h. 10)60
i. 1)10
j. 9)54

k. 5)35
l. 4)36
m. 3)9
n. 9)18
o. 3)12

p. 6)12
q. 3)18
r. 9)27
s. 8)40
t. 7)7

u. 10)30
v. 3)15
w. 6)54
x. 2)16
y. 4)20

z. 2)20
A. 9)36
B. 5)25
C. 5)5
D. 5)45

Lesson

a. ____ ⌐4→ 16   d. 9 ⌐→ 45   g. 3 ⌐→ 12   j. 6 ⌐→ 36

b. 6 ⌐6→ ____   e. 9 ⌐4→ ____   h. ____ ⌐4→ 24   k. 9 ⌐→ 27

c. 6 ⌐→ 24   f. 3 ⌐6→ ____   i. ____ ⌐3→ 9   l. ____ ⌐6→ 54

Part 2

a.	b.	c.	d.	e.
92	35	51	46	93
× 4	× 9	× 7	× 5	× 3

Part 3

a. 4⟌16   f. 4⟌36   k. 9⟌54   p. 3⟌27   u. 8⟌80

b. 6⟌36   g. 9⟌45   l. 5⟌40   q. 6⟌54   v. 6⟌24

c. 4⟌24   h. 4⟌12   m. 6⟌24   r. 9⟌36   w. 5⟌25

d. 3⟌6   i. 6⟌12   n. 5⟌30   s. 7⟌35   x. 9⟌45

e. 3⟌18   j. 3⟌9   o. 2⟌6   t. 9⟌27   y. 4⟌20

Part 4

a.	b.	c.	d.
927	614	365	412
−189	−587	− 96	−248

Part 5

a. 6 × 6 =   g. 3 × 9 =   m. 1 × 7 =   s. 0 × 5 =   y. 5 × 8 =

b. 3 × 4 =   h. 7 × 5 =   n. 2 × 8 =   t. 9 × 2 =   z. 6 × 4 =

c. 5 × 5 =   i. 6 × 3 =   o. 4 × 3 =   u. 5 × 6 =   A. 5 × 2 =

d. 9 × 4 =   j. 4 × 1 =   p. 7 × 10 =   v. 9 × 3 =   B. 9 × 5 =

e. 4 × 6 =   k. 9 × 6 =   q. 4 × 9 =   w. 3 × 0 =   C. 10 × 4 =

f. 3 × 3 =   l. 4 × 4 =   r. 3 × 6 =   x. 6 × 9 =   D. 3 × 2 =

Lesson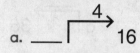

a. ___⌐4→ 16   d. 3⌐→ 9   g. 4⌐4→ ___   j. ___⌐3→ 27

b. 4⌐=→ 12   e. 6⌐=→ 36   h. ___⌐4→ 36   k. 9⌐=→ 54

c. ___⌐4→ 24   f. 6⌐3→ ___   i. 6⌐=→ 24   l. 4⌐3→ ___

Part 2

a. 6⟌36   g. 3⟌18   m. 4⟌24   s. 9⟌36   y. 9⟌45

b. 4⟌36   h. 2⟌18   n. 5⟌25   t. 6⟌36   z. 10⟌80

c. 4⟌12   i. 6⟌24   o. 3⟌6   u. 3⟌3   A. 6⟌30

d. 2⟌12   j. 3⟌9   p. 4⟌20   v. 2⟌10   B. 6⟌54

e. 9⟌54   k. 9⟌0   q. 6⟌18   w. 3⟌27   C. 4⟌20

f. 5⟌5   l. 4⟌16   r. 9⟌27   x. 3⟌12   D. 7⟌35

Part 3

a. $6 \times 4 =$ g. $3 \times 3 =$ m. $4 \times 10 =$ s. $3 \times 6 =$ y. $9 \times 3 =$

b. $3 \times 9 =$ h. $9 \times 4 =$ n. $5 \times 5 =$ t. $10 \times 10 =$ z. $2 \times 5 =$

c. $6 \times 6 =$ i. $6 \times 3 =$ o. $4 \times 6 =$ u. $2 \times 7 =$ A. $1 \times 3 =$

d. $4 \times 4 =$ j. $1 \times 7 =$ p. $7 \times 5 =$ v. $6 \times 0 =$ B. $5 \times 9 =$

e. $0 \times 8 =$ k. $3 \times 4 =$ q. $8 \times 1 =$ w. $4 \times 3 =$ C. $4 \times 2 =$

f. $6 \times 9 =$ l. $5 \times 8 =$ r. $4 \times 9 =$ x. $6 \times 5 =$ D. $9 \times 6 =$

Lesson

Part 1

a. ___ ⌐ 3→ 9   d. 6 ⌐⇒ 36   g. 6 ⌐⇒ 24   j. ___ ⌐ 4→ 12

b. 6 ⌐⇒ 18   e. 4 ⌐ 3→ ___   h. 9 ⌐ 3→ ___   k. 6 ⌐ 4→ ___

c. ___ ⌐ 4→ 16   f. 9 ⌐⇒ 54   i. ___ ⌐ 6→ 12   l. ___ ⌐ 3→ 18

Part 2

a. 3⟌27   g. 4⟌12   m. 9⟌54   s. 2⟌14   y. 7⟌70

b. 4⟌16   h. 5⟌40   n. 3⟌12   t. 6⟌36   z. 6⟌30

c. 6⟌18   i. 7⟌0   o. 6⟌24   u. 4⟌36   A. 5⟌35

d. 9⟌27   j. 9⟌36   p. 4⟌24   v. 5⟌45   B. 5⟌20

e. 6⟌24   k. 5⟌30   q. 2⟌18   w. 10⟌80   C. 10⟌60

f. 6⟌54   l. 3⟌18   r. 9⟌27   x. 1⟌5   D. 6⟌12

Part 3

a. $\dfrac{6}{7} + \dfrac{4}{7} =$ b. $\dfrac{18}{5} - \dfrac{10}{5} =$ c. $\dfrac{17}{4} - \dfrac{2}{4} =$

Part 4

a. $3 \times 3 =$ g. $9 \times 3 =$ m. $6 \times 4 =$ s. $4 \times 3 =$ y. $9 \times 6 =$

b. $4 \times 9 =$ h. $6 \times 6 =$ n. $3 \times 9 =$ t. $7 \times 5 =$ z. $8 \times 0 =$

c. $1 \times 8 =$ i. $7 \times 2 =$ o. $5 \times 8 =$ u. $2 \times 10 =$ A. $3 \times 4 =$

d. $6 \times 3 =$ j. $4 \times 4 =$ p. $7 \times 1 =$ v. $9 \times 4 =$ B. $2 \times 5 =$

e. $5 \times 5 =$ k. $3 \times 6 =$ q. $3 \times 4 =$ w. $5 \times 3 =$ C. $10 \times 7 =$

f. $9 \times 2 =$ l. $0 \times 7 =$ r. $6 \times 9 =$ x. $8 \times 2 =$ D. $4 \times 6 =$

Lesson

Part 1

a. 6 ⟶³ ___

b. ___ ⟶⁴ 16

c. ___ ⟶⁴ 36

d. 6 ⟹ 36

e. 3 ⟶³ ___

f. 9 ⟹ 18

g. ___ ⟶³ 18

h. 4 ⟹ 12

i. ___ ⟶² 12

j. 9 ⟹ 27

k. ___ ⟶⁶ 54

l. 6 ⟶⁶ ___

Part 2

a. 4$\overline{)16}$ g. 9$\overline{)36}$ m. 3$\overline{)9}$ s. 4$\overline{)12}$ y. 5$\overline{)30}$

b. 6$\overline{)24}$ h. 6$\overline{)36}$ n. 3$\overline{)27}$ t. 2$\overline{)12}$ z. 3$\overline{)30}$

c. 4$\overline{)24}$ i. 1$\overline{)1}$ o. 2$\overline{)4}$ u. 2$\overline{)14}$ A. 4$\overline{)4}$

d. 3$\overline{)12}$ j. 7$\overline{)0}$ p. 2$\overline{)8}$ v. 6$\overline{)36}$ B. 4$\overline{)0}$

e. 6$\overline{)12}$ k. 9$\overline{)18}$ q. 6$\overline{)54}$ w. 4$\overline{)36}$ C. 6$\overline{)18}$

f. 1$\overline{)9}$ l. 3$\overline{)18}$ r. 9$\overline{)54}$ x. 10$\overline{)60}$ D. 5$\overline{)35}$

Part 3

a. $\begin{array}{r} \$16.37 \\ -4.44 \\ \hline \end{array}$

c. $\begin{array}{r} \$19.87 \\ -12.59 \\ \hline \end{array}$

b. $\begin{array}{r} \$45.08 \\ +16.07 \\ \hline \end{array}$

Part 4

a. $\dfrac{7}{9} - \dfrac{7}{9} =$ c. $\dfrac{8}{9} + \dfrac{8}{9} =$

b. $\dfrac{3}{5} + \dfrac{13}{5} =$ d. $\dfrac{14}{8} - \dfrac{6}{8} =$

Connecting Math Concepts

Lesson 29

Part 5

a. 6 × 4 =	g. 9 × 6 =	m. 2 × 2 =	s. 5 × 9 =	y. 5 × 7 =
b. 4 × 9 =	h. 3 × 3 =	n. 6 × 9 =	t. 0 × 4 =	z. 4 × 10 =
c. 3 × 6 =	i. 6 × 6 =	o. 3 × 4 =	u. 9 × 4 =	A. 1 × 6 =
d. 5 × 4 =	j. 4 × 1 =	p. 6 × 5 =	v. 6 × 2 =	B. 4 × 3 =
e. 6 × 0 =	k. 9 × 3 =	q. 4 × 6 =	w. 8 × 5 =	C. 2 × 9 =
f. 4 × 4 =	l. 10 × 6 =	r. 2 × 4 =	x. 6 × 3 =	D. 5 × 5 =

Lesson 30

Part 1

a. 6 × 6 =	g. 5 × 5 =	m. 2 × 10 =	s. 9 × 2 =	y. 5 × 6 =
b. 4 × 9 =	h. 9 × 3 =	n. 4 × 5 =	t. 0 × 8 =	z. 9 × 4 =
c. 6 × 9 =	i. 6 × 4 =	o. 3 × 4 =	u. 5 × 8 =	A. 3 × 2 =
d. 3 × 3 =	j. 4 × 0 =	p. 2 × 6 =	v. 10 × 4 =	B. 1 × 3 =
e. 4 × 4 =	k. 4 × 3 =	q. 9 × 6 =	w. 4 × 6 =	C. 5 × 4 =
f. 2 × 8 =	l. 6 × 3 =	r. 3 × 6 =	x. 8 × 1 =	D. 3 × 9 =

Part 2

a. $84.07 − 54.65	b. $ 9.58 +23.91	c. $36.15 − 7.09	d. $32.57 +16.50

Part 3

a. 3⟌27	g. 2⟌18	m. 1⟌3	s. 8⟌8	y. 9⟌45
b. 6⟌36	h. 3⟌18	n. 4⟌40	t. 10⟌20	z. 6⟌12
c. 6⟌54	i. 3⟌9	o. 5⟌40	u. 5⟌20	A. 7⟌0
d. 4⟌24	j. 9⟌0	p. 9⟌27	v. 6⟌18	B. 6⟌24
e. 8⟌16	k. 9⟌36	q. 2⟌12	w. 5⟌30	C. 9⟌54
f. 4⟌16	l. 6⟌36	r. 4⟌12	x. 3⟌12	D. 6⟌30

Lesson

Part 1

a. ___ ⌐4→ 36   d. 9 ⌐9→ ___   g. 6 ⌐→ 36   j. 9 ⌐→ 72

b. 6 ⌐6→ ___   e. ___ ⌐4→ 24   h. 9 ⌐8→ ___   k. 9 ⌐→ 63

c. ___ ⌐3→ 12   f. 9 ⌐7→ ___   i. ___ ⌐4→ 16   l. 9 ⌐→ 81

Part 2

a. 9⟌81   g. 8⟌16   m. 4⟌24   s. 6⟌18   y. 9⟌27

b. 6⟌36   h. 8⟌72   n. 2⟌18   t. 4⟌36   z. 4⟌28

c. 9⟌45   i. 9⟌54   o. 3⟌27   u. 4⟌32   A. 7⟌49

d. 9⟌72   j. 7⟌63   p. 3⟌18   v. 5⟌20   B. 8⟌32

e. 9⟌36   k. 3⟌12   q. 6⟌24   w. 5⟌25   C. 4⟌20

f. 9⟌63   l. 4⟌16   r. 6⟌54   x. 7⟌28   D. 8⟌64

Part 3

a. $\dfrac{18}{2} + \dfrac{1}{3} =$ c. $\dfrac{15}{4} - \dfrac{2}{3} =$ e. $\dfrac{10}{9} - \dfrac{3}{9} =$

b. $\dfrac{5}{3} - \dfrac{2}{3} =$ d. $\dfrac{3}{4} + \dfrac{9}{4} =$ f. $\dfrac{1}{6} + \dfrac{1}{8} =$

Part 4

a. 251
  − 179

b. 417
  − 54

c. 682
  − 159

d. 456
  − 98

Lesson 31

Part 5

a. $4 \times 4 =$ g. $10 \times 0 =$ m. $5 \times 5 =$ s. $6 \times 3 =$ y. $4 \times 10 =$

b. $9 \times 9 =$ h. $4 \times 3 =$ n. $2 \times 8 =$ t. $7 \times 5 =$ z. $6 \times 9 =$

c. $6 \times 6 =$ i. $7 \times 9 =$ o. $3 \times 9 =$ u. $9 \times 6 =$ A. $5 \times 8 =$

d. $9 \times 4 =$ j. $3 \times 3 =$ p. $4 \times 6 =$ v. $3 \times 4 =$ B. $9 \times 3 =$

e. $3 \times 6 =$ k. $6 \times 4 =$ q. $8 \times 9 =$ w. $4 \times 6 =$ C. $2 \times 6 =$

f. $9 \times 8 =$ l. $1 \times 7 =$ r. $7 \times 2 =$ x. $9 \times 7 =$ D. $4 \times 9 =$

Lesson 32

Part 1

a. ___ ⌐6→ 54   d. ___ ⌐9→ 81   g. 9 ⌐8→ ___   j. ___ ⌐8→ 72

b. 6 ⌐4→ ___   e. ___ ⌐6→ 36   h. 9 ⌐→ 63   k. 3 ⌐→ 18

c. 9 ⌐→ 36   f. 4 ⌐→ 16   i. 9 ⌐→ 27   l. ___ ⌐7→ 63

Part 2

a. $6\overline{)36}$ g. $4\overline{)24}$ m. $3\overline{)27}$ s. $7\overline{)35}$ y. $7\overline{)63}$

b. $5\overline{)35}$ h. $4\overline{)36}$ n. $4\overline{)16}$ t. $6\overline{)24}$ z. $4\overline{)20}$

c. $6\overline{)30}$ i. $8\overline{)80}$ o. $9\overline{)36}$ u. $4\overline{)12}$ A. $9\overline{)27}$

d. $9\overline{)72}$ j. $9\overline{)54}$ p. $8\overline{)72}$ v. $10\overline{)80}$ B. $6\overline{)54}$

e. $3\overline{)18}$ k. $3\overline{)12}$ q. $2\overline{)4}$ w. $9\overline{)81}$ C. $5\overline{)20}$

f. $9\overline{)63}$ l. $5\overline{)25}$ r. $6\overline{)18}$ x. $5\overline{)30}$ D. $2\overline{)18}$

Lesson 32

a. $\dfrac{3}{4} + \dfrac{9}{5} =$ c. $\dfrac{10}{9} + \dfrac{10}{9} =$ e. $\dfrac{6}{8} - \dfrac{1}{8} =$

b. $\dfrac{15}{7} - \dfrac{2}{7} =$ d. $\dfrac{11}{8} - \dfrac{11}{9} =$ f. $\dfrac{9}{7} + \dfrac{7}{2} =$

Part 4

a. $6 \times 3 =$	g. $6 \times 4 =$	m. $4 \times 6 =$	s. $7 \times 5 =$	y. $8 \times 9 =$
b. $4 \times 4 =$	h. $9 \times 9 =$	n. $7 \times 1 =$	t. $9 \times 4 =$	z. $3 \times 6 =$
c. $9 \times 8 =$	i. $3 \times 3 =$	o. $3 \times 9 =$	u. $3 \times 10 =$	A. $7 \times 0 =$
d. $3 \times 4 =$	j. $9 \times 6 =$	p. $8 \times 2 =$	v. $6 \times 2 =$	B. $10 \times 2 =$
e. $6 \times 6 =$	k. $0 \times 5 =$	q. $4 \times 3 =$	w. $9 \times 7 =$	C. $5 \times 4 =$
f. $7 \times 9 =$	l. $4 \times 9 =$	r. $6 \times 9 =$	x. $5 \times 8 =$	D. $9 \times 3 =$

Lesson 33

Part 1

a. ___ $\xrightarrow{\ 3\ }$ 9   d. 9 $\xrightarrow{\ 6\ }$ ___   g. 9 $\xrightarrow{\ 4\ }$ ___   j. 6 $\xrightarrow{\ 6\ }$ ___

b. 6 $\xrightarrow{\quad}$ 18   e. 4 $\xrightarrow{\quad}$ 12   h. ___ $\xrightarrow{\ 3\ }$ 27   k. ___ $\xrightarrow{\ 2\ }$ 16

c. 9 $\xrightarrow{\quad}$ 81   f. ___ $\xrightarrow{\ 8\ }$ 72   i. ___ $\xrightarrow{\ 4\ }$ 36   l. 9 $\xrightarrow{\ 7\ }$ ___

Lesson 33

Part 2

a. $6\overline{)36}$ g. $8\overline{)72}$ m. $5\overline{)35}$ s. $9\overline{)27}$ y. $6\overline{)18}$

b. $9\overline{)45}$ h. $9\overline{)54}$ n. $4\overline{)24}$ t. $5\overline{)30}$ z. $4\overline{)20}$

c. $9\overline{)72}$ i. $7\overline{)63}$ o. $5\overline{)45}$ u. $3\overline{)18}$ A. $10\overline{)80}$

d. $9\overline{)36}$ j. $3\overline{)12}$ p. $1\overline{)5}$ v. $7\overline{)35}$ B. $4\overline{)36}$

e. $9\overline{)63}$ k. $4\overline{)16}$ q. $8\overline{)80}$ w. $5\overline{)30}$ C. $5\overline{)25}$

f. $8\overline{)16}$ l. $6\overline{)54}$ r. $4\overline{)12}$ x. $3\overline{)27}$ D. $6\overline{)24}$

Part 3

a. $5\overline{)37}$ c. $5\overline{)32}$ e. $5\overline{)18}$

b. $5\overline{)14}$ d. $5\overline{)26}$ f. $5\overline{)49}$

Part 4

a. $\dfrac{26}{5} - \dfrac{5}{3} =$ c. $\dfrac{6}{10} - \dfrac{6}{10} =$ e. $\dfrac{12}{7} + \dfrac{5}{7} =$

b. $\dfrac{18}{3} + \dfrac{2}{3} =$ d. $\dfrac{8}{4} - \dfrac{8}{3} =$ f. $\dfrac{10}{3} + \dfrac{3}{8} =$

Lesson 33

Part 5

a. $6 \times 9 =$ g. $9 \times 4 =$ m. $3 \times 9 =$ s. $3 \times 4 =$ y. $8 \times 9 =$

b. $3 \times 3 =$ h. $6 \times 6 =$ n. $5 \times 4 =$ t. $6 \times 2 =$ z. $10 \times 3 =$

c. $2 \times 8 =$ i. $4 \times 3 =$ o. $9 \times 6 =$ u. $9 \times 7 =$ A. $4 \times 2 =$

d. $4 \times 4 =$ j. $7 \times 9 =$ p. $8 \times 0 =$ v. $1 \times 10 =$ B. $5 \times 8 =$

e. $9 \times 9 =$ k. $4 \times 6 =$ q. $9 \times 2 =$ w. $4 \times 9 =$ C. $9 \times 3 =$

f. $3 \times 6 =$ l. $9 \times 8 =$ r. $6 \times 3 =$ x. $5 \times 5 =$ D. $6 \times 4 =$

Lesson 34

Part 1

a. $6 \xrightarrow{\quad} 36$ d. $6 \xrightarrow{\quad} 24$ g. $3 \xrightarrow{9} \_\_$ j. $\_\_ \xrightarrow{6} 54$

b. $8 \xrightarrow{2} \_\_$ e. $9 \xrightarrow{9} \_\_$ h. $\_\_ \xrightarrow{3} 9$ k. $4 \xrightarrow{4} \_\_$

c. $\_\_ \xrightarrow{3} 12$ f. $6 \xrightarrow{\quad} 12$ i. $\_\_ \xrightarrow{2} 18$ l. $\_\_ \xrightarrow{7} 63$

Part 2

a. $3 \overline{)9}$ g. $9 \overline{)54}$ m. $6 \overline{)36}$ s. $5 \overline{)30}$ y. $9 \overline{)45}$

b. $4 \overline{)12}$ h. $9 \overline{)81}$ n. $3 \overline{)12}$ t. $3 \overline{)27}$ z. $6 \overline{)54}$

c. $4 \overline{)16}$ i. $6 \overline{)24}$ o. $9 \overline{)72}$ u. $3 \overline{)18}$ A. $5 \overline{)20}$

d. $9 \overline{)63}$ j. $6 \overline{)18}$ p. $4 \overline{)24}$ v. $7 \overline{)63}$ B. $5 \overline{)25}$

e. $9 \overline{)27}$ k. $9 \overline{)45}$ q. $8 \overline{)72}$ w. $2 \overline{)16}$ C. $1 \overline{)5}$

f. $5 \overline{)35}$ l. $10 \overline{)80}$ r. $2 \overline{)10}$ x. $5 \overline{)45}$ D. $9 \overline{)36}$

Lesson 34

Part 3

a. $\dfrac{7}{3} + \dfrac{7}{10} =$ c. $\dfrac{14}{1} + \dfrac{5}{1} =$ e. $\dfrac{16}{5} + \dfrac{10}{5} =$

b. $\dfrac{10}{7} - \dfrac{3}{7} =$ d. $\dfrac{16}{5} - \dfrac{10}{16} =$ f. $\dfrac{16}{5} - \dfrac{16}{5} =$

Part 4

a. $5\overline{)8}$ c. $3\overline{)19}$ e. $5\overline{)47}$

b. $4\overline{)11}$ d. $5\overline{)31}$ f. $2\overline{)17}$

Part 5

a. $4 \times 6 =$	g. $8 \times 9 =$	m. $3 \times 9 =$	s. $4 \times 3 =$	y. $9 \times 8 =$
b. $3 \times 4 =$	h. $6 \times 6 =$	n. $5 \times 4 =$	t. $9 \times 6 =$	z. $5 \times 5 =$
c. $9 \times 9 =$	i. $9 \times 4 =$	o. $2 \times 10 =$	u. $7 \times 2 =$	A. $10 \times 10 =$
d. $7 \times 0 =$	j. $3 \times 3 =$	p. $8 \times 5 =$	v. $4 \times 9 =$	B. $6 \times 4 =$
e. $4 \times 4 =$	k. $1 \times 8 =$	q. $6 \times 9 =$	w. $5 \times 8 =$	C. $7 \times 9 =$
f. $6 \times 3 =$	l. $9 \times 7 =$	r. $4 \times 2 =$	x. $3 \times 6 =$	D. $9 \times 3 =$

Lesson 35

Part 1

a. $9 \xrightarrow{\ 7\ } \underline{\quad}$ d. $\underline{\quad} \xrightarrow{\ 3\ } 18$ g. $6 \xrightarrow{\ 4\ } \underline{\quad}$ j. $\underline{\quad} \xrightarrow{\ 4\ } 36$

b. $7 \xrightarrow{\ 2\ } \underline{\quad}$ e. $5 \xrightarrow{\ 7\ } \underline{\quad}$ h. $\underline{\quad} \xrightarrow{\ 7\ } 63$ k. $9 \xrightarrow{\quad} 54$

c. $\underline{\quad} \xrightarrow{\ 2\ } 18$ f. $\underline{\quad} \xrightarrow{\ 8\ } 72$ i. $\underline{\quad} \xrightarrow{\ 4\ } 16$ l. $4 \xrightarrow{\ 3\ } \underline{\quad}$

Lesson 35

Part 2

a. $6\overline{\smash{)}36}$	g. $8\overline{\smash{)}72}$	m. $4\overline{\smash{)}24}$	s. $4\overline{\smash{)}12}$	y. $10\overline{\smash{)}80}$
b. $6\overline{\smash{)}18}$	h. $9\overline{\smash{)}36}$	n. $3\overline{\smash{)}9}$	t. $9\overline{\smash{)}54}$	z. $5\overline{\smash{)}45}$
c. $9\overline{\smash{)}27}$	i. $9\overline{\smash{)}45}$	o. $6\overline{\smash{)}24}$	u. $5\overline{\smash{)}35}$	A. $9\overline{\smash{)}72}$
d. $2\overline{\smash{)}10}$	j. $2\overline{\smash{)}16}$	p. $4\overline{\smash{)}16}$	v. $2\overline{\smash{)}12}$	B. $6\overline{\smash{)}54}$
e. $5\overline{\smash{)}5}$	k. $5\overline{\smash{)}10}$	q. $5\overline{\smash{)}20}$	w. $9\overline{\smash{)}81}$	C. $7\overline{\smash{)}35}$
f. $5\overline{\smash{)}25}$	l. $5\overline{\smash{)}30}$	r. $3\overline{\smash{)}27}$	x. $4\overline{\smash{)}36}$	D. $6\overline{\smash{)}30}$

Part 3

a. __ 3   c. __ 10

b. __ 9   d. __ 5

Part 4

a. $3\overline{\smash{)}10}$ c. $5\overline{\smash{)}19}$ e. $3\overline{\smash{)}29}$

b. $4\overline{\smash{)}10}$ d. $2\overline{\smash{)}19}$ f. $5\overline{\smash{)}29}$

Part 5

a. $6 \times 6 =$	g. $3 \times 4 =$	m. $3 \times 9 =$	s. $9 \times 2 =$	y. $1 \times 5 =$
b. $9 \times 4 =$	h. $8 \times 9 =$	n. $2 \times 4 =$	t. $3 \times 6 =$	z. $7 \times 9 =$
c. $3 \times 3 =$	i. $4 \times 6 =$	o. $6 \times 9 =$	u. $10 \times 7 =$	A. $6 \times 4 =$
d. $9 \times 9 =$	j. $0 \times 10 =$	p. $5 \times 7 =$	v. $9 \times 8 =$	B. $9 \times 3 =$
e. $2 \times 8 =$	k. $6 \times 3 =$	q. $8 \times 1 =$	w. $10 \times 3 =$	C. $8 \times 5 =$
f. $4 \times 4 =$	l. $9 \times 7 =$	r. $4 \times 3 =$	x. $4 \times 9 =$	D. $9 \times 6 =$

Connecting Math Concepts

Lesson 36

Part 1

a. 3 ⌐7→ ___   d. ___ ⌐6→ 54   g. 3 ⌐→ 24   j. ___ ⌐3→ 12

b. 5 ⌐→ 35   e. ___ ⌐2→ 18   h. ___ ⌐3→ 18   k. ___ ⌐4→ 36

c. ___ ⌐7→ 63   f. 9 ⌐→ 54   i. 6 ⌐4→ ___   l. ___ ⌐7→ 21

Part 2

a. 10$\overline{)80}$ f. 6$\overline{)18}$ k. 9$\overline{)27}$ p. 3$\overline{)21}$ u. 5$\overline{)5}$ z. 1$\overline{)8}$

b. 4$\overline{)12}$ g. 9$\overline{)36}$ l. 9$\overline{)45}$ q. 2$\overline{)16}$ v. 6$\overline{)18}$ A. 7$\overline{)21}$

c. 4$\overline{)24}$ h. 3$\overline{)9}$ m. 6$\overline{)24}$ r. 4$\overline{)16}$ w. 5$\overline{)20}$ B. 3$\overline{)27}$

d. 8$\overline{)72}$ i. 9$\overline{)54}$ n. 8$\overline{)24}$ s. 2$\overline{)12}$ x. 9$\overline{)81}$ C. 4$\overline{)36}$

e. 6$\overline{)36}$ j. 5$\overline{)45}$ o. 9$\overline{)72}$ t. 6$\overline{)54}$ y. 7$\overline{)35}$ D. 6$\overline{)30}$

Part 3

a. 4$\overline{)16}$ d. 5$\overline{)10}$

b. 5$\overline{)16}$ e. 5$\overline{)27}$

c. 4$\overline{)10}$ f. 3$\overline{)27}$

Part 4

	Monday	Tuesday	Total for both days
Red birds	24		41
Yellow birds	13		
Total birds		57	

a. How many yellow birds were seen on Tuesday? _____

b. How many yellow birds were seen on both days? _____

c. How many red birds were seen on Monday? _____

d. Were more birds seen on Monday or Tuesday?

_____

Lesson 36

Part 5

a. $7 \times 3 =$ g. $4 \times 6 =$ m. $3 \times 7 =$ s. $8 \times 3 =$ y. $3 \times 6 =$

b. $9 \times 9 =$ h. $3 \times 3 =$ n. $6 \times 9 =$ t. $9 \times 4 =$ z. $10 \times 4 =$

c. $3 \times 8 =$ i. $1 \times 9 =$ o. $3 \times 4 =$ u. $7 \times 0 =$ A. $9 \times 3 =$

d. $9 \times 6 =$ j. $6 \times 6 =$ p. $2 \times 6 =$ v. $6 \times 4 =$ B. $8 \times 9 =$

e. $4 \times 3 =$ k. $9 \times 8 =$ q. $9 \times 7 =$ w. $3 \times 9 =$ C. $7 \times 5 =$

f. $7 \times 9 =$ l. $4 \times 4 =$ r. $6 \times 3 =$ x. $5 \times 8 =$ D. $4 \times 9 =$

Lesson 37

Part 1

a. $9 \longrightarrow 18$ d. $9 \longrightarrow 54$ g. $9 \longrightarrow 36$ j. $\underline{\quad} \xrightarrow{7} 63$

b. $\underline{\quad} \xrightarrow{4} 24$ e. $6 \longrightarrow 18$ h. $3 \longrightarrow 21$ k. $5 \xrightarrow{7} \underline{\quad}$

c. $3 \xrightarrow{7} \underline{\quad}$ f. $\underline{\quad} \xrightarrow{6} 54$ i. $4 \longrightarrow 12$ l. $\underline{\quad} \xrightarrow{2} 14$

Part 2

a. $9\overline{)81}$ f. $2\overline{)12}$ k. $3\overline{)27}$ p. $9\overline{)72}$ u. $8\overline{)72}$ z. $9\overline{)45}$

b. $5\overline{)20}$ g. $7\overline{)35}$ l. $5\overline{)30}$ q. $2\overline{)16}$ v. $4\overline{)24}$ A. $3\overline{)9}$

c. $5\overline{)10}$ h. $6\overline{)30}$ m. $4\overline{)16}$ r. $2\overline{)10}$ w. $10\overline{)80}$ B. $9\overline{)27}$

d. $5\overline{)5}$ i. $5\overline{)25}$ n. $6\overline{)24}$ s. $6\overline{)18}$ x. $4\overline{)12}$ C. $9\overline{)54}$

e. $6\overline{)54}$ j. $4\overline{)36}$ o. $5\overline{)35}$ t. $6\overline{)36}$ y. $9\overline{)36}$ D. $5\overline{)45}$

Lesson 37

Part 3

_____

_____

_____

	Hardcover books	Softcover books	Total books
Library		2046	5359
Bookstore	1150		3540
Total for both places			

a. How many total books were in both places?

_____

b. How many softcover books were in the library?

_____

c. Were there fewer softcover books in the library or the bookstore? _____

d. Were there more hardcover or softcover books in the bookstore? _____

e. Were there more total books in the library or the bookstore? _____

Part 4

a. $3 \times 6 =$

b. $4 \times 4 =$

c. $6 \times 9 =$

d. $7 \times 3 =$

e. $9 \times 9 =$

f. $6 \times 4 =$

g. $4 \times 9 =$

h. $8 \times 3 =$

i. $9 \times 7 =$

j. $6 \times 6 =$

k. $3 \times 3 =$

l. $8 \times 9 =$

m. $3 \times 9 =$

n. $4 \times 3 =$

o. $0 \times 5 =$

p. $9 \times 8 =$

q. $3 \times 7 =$

r. $4 \times 6 =$

s. $6 \times 3 =$

t. $7 \times 9 =$

u. $8 \times 5 =$

v. $3 \times 4 =$

w. $9 \times 6 =$

x. $10 \times 7 =$

y. $9 \times 4 =$

z. $3 \times 8 =$

A. $5 \times 6 =$

B. $3 \times 10 =$

C. $8 \times 1 =$

D. $9 \times 3 =$

Lesson

Part 1

a. 5 ⌐7→ ___

b. 9 ⌐→ 63

c. 6 ⌐→ 18

d. ___ ⌐3→ 21

e. 9 ⌐→ 36

f. 9 ⌐→ 54

g. 9 ⌐→ 18

h. 4 ⌐→ 24

i. 4 ⌐→ 12

j. ___ ⌐2→ 14

k. 6 ⌐→ 54

l. 9 ⌐7→ ___

Part 2

a. 4⟌12   f. 6⟌18   k. 6⟌24   p. 5⟌45   u. 5⟌5   z. 7⟌35

b. 6⟌60   g. 4⟌16   l. 2⟌10   q. 3⟌21   v. 5⟌10   A. 3⟌27

c. 4⟌24   h. 2⟌16   m. 3⟌9   r. 9⟌45   w. 9⟌81   B. 8⟌40

d. 8⟌24   i. 9⟌72   n. 9⟌27   s. 2⟌12   x. 5⟌20   C. 3⟌24

e. 6⟌36   j. 5⟌35   o. 9⟌54   t. 6⟌54   y. 9⟌36   D. 5⟌25

Part 3

_____

_____

_____

_____

	North	South	Total in both directions
Mountain Highway		463	770
Valley Parkway		458	782
Total for both roads			

a. How many vehicles on both roads were traveling north? _____

b. Were there fewer vehicles traveling south on Mountain Highway or Valley Parkway?

_____

c. Were there more vehicles traveling north or south on Mountain Highway? _____

d. 324 vehicles were traveling _____ on _____.

e. 770 vehicles were traveling _____ on _____.

Lesson 38

Part 4

a. 3 × 7 =	g. 4 × 6 =	m. 4 × 9 =	s. 8 × 9 =	y. 9 × 6 =
b. 6 × 9 =	h. 7 × 3 =	n. 3 × 8 =	t. 6 × 4 =	z. 5 × 10 =
c. 3 × 3 =	i. 3 × 4 =	o. 7 × 9 =	u. 9 × 3 =	A. 6 × 3 =
d. 4 × 4 =	j. 6 × 6 =	p. 0 × 6 =	v. 10 × 1 =	B. 2 × 4 =
e. 8 × 3 =	k. 9 × 9 =	q. 4 × 3 =	w. 9 × 4 =	C. 3 × 9 =
f. 9 × 7 =	l. 9 × 8 =	r. 5 × 7 =	x. 3 × 6 =	D. 8 × 2 =

Lesson 39

Part 1

a. 3 ⌐7→ ___   d. ___ ⌐2→ 18   g. 9 ⌐8→ ___   j. 7 ⌐=→ 21

b. 9 ⌐→ 63   e. ___ ⌐7→ 35   h. 8 ⌐=→ 24   k. 9 ⌐→ 54

c. ___ ⌐4→ 36   f. ___ ⌐3→ 18   i. 6 ⌐=→ 24   l. ___ ⌐4→ 24

Part 2

a. 3⟌21	g. 1⟌6	m. 9⟌54	s. 9⟌72	y. 4⟌24
b. 6⟌30	h. 5⟌10	n. 5⟌45	t. 5⟌35	z. 8⟌72
c. 5⟌25	i. 4⟌20	o. 9⟌45	u. 4⟌16	A. 6⟌18
d. 4⟌36	j. 9⟌81	p. 8⟌24	v. 6⟌24	B. 6⟌36
e. 10⟌80	k. 8⟌16	q. 9⟌27	w. 6⟌54	C. 7⟌21
f. 3⟌24	l. 2⟌10	r. 2⟌12	x. 4⟌12	D. 5⟌15

Lesson 39

Part 3

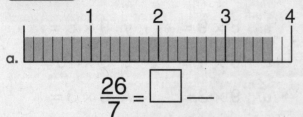

a. $\frac{26}{7} = \boxed{} \underline{\quad}$

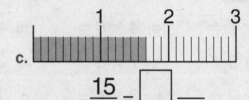

c. $\frac{15}{9} = \boxed{} \underline{\quad}$

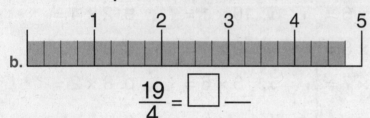

b. $\frac{19}{4} = \boxed{} \underline{\quad}$

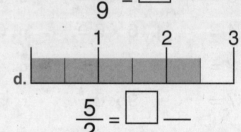

d. $\frac{5}{2} = \boxed{} \underline{\quad}$

Part 4

_____

_____

_____

_____

	Cars	Trucks	Total vehicles
Airport lot	352	297	
Green Street lot		413	
Total for both lots	606		

a. How many trucks were parked in both lots?

_____

b. Were there more cars or trucks in the Green Street lot? _____

c. Were there fewer total vehicles in Airport lot or Green Street lot? _____

d. 649 _____ were parked in _____.

e. 254 _____ were parked in _____.

Part 5

a. 8 ⟌ 4 3 ___

b. 9 ⟌ 2 9 ___

c. 4 ⟌ 1 1 ___

d. 9 ⟌ 6 8 ___

e. 4 ⟌ 1 4 ___

f. 5 ⟌ 3 4 ___

Lesson 39

Part 6

a. $8 \times 9 =$ g. $3 \times 4 =$ m. $4 \times 9 =$ s. $4 \times 0 =$ y. $4 \times 3 =$

b. $4 \times 4 =$ h. $9 \times 7 =$ n. $3 \times 6 =$ t. $9 \times 8 =$ z. $2 \times 2 =$

c. $9 \times 3 =$ i. $3 \times 8 =$ o. $2 \times 10 =$ u. $3 \times 7 =$ A. $3 \times 9 =$

d. $4 \times 6 =$ j. $6 \times 6 =$ p. $6 \times 9 =$ v. $9 \times 6 =$ B. $5 \times 4 =$

e. $7 \times 3 =$ k. $1 \times 9 =$ q. $8 \times 3 =$ w. $6 \times 4 =$ C. $6 \times 3 =$

f. $9 \times 9 =$ l. $3 \times 3 =$ r. $9 \times 4 =$ x. $7 \times 9 =$ D. $9 \times 2 =$

Lesson 40

Part 1

a. 6 ⟶ 24   d. ⟶6 54   g. ⟶4 12   j. 9 ⟶ 72

b. 9 ⟶6 ___   e. 9 ⟶ 63   h. ⟶7 35   k. ⟶8 24

c. 3 ⟶7 ___   f. ⟶4 36   i. ⟶6 18   l. 4 ⟶ 16

Part 2

a. $8\overline{)72}$ g. $5\overline{)20}$ m. $9\overline{)36}$ s. $6\overline{)54}$ y. $3\overline{)9}$

b. $7\overline{)35}$ h. $9\overline{)45}$ n. $9\overline{)81}$ t. $7\overline{)7}$ z. $4\overline{)24}$

c. $8\overline{)24}$ i. $4\overline{)16}$ o. $4\overline{)36}$ u. $5\overline{)15}$ A. $9\overline{)72}$

d. $5\overline{)10}$ j. $6\overline{)18}$ p. $10\overline{)80}$ v. $4\overline{)12}$ B. $9\overline{)54}$

e. $6\overline{)30}$ k. $6\overline{)36}$ q. $2\overline{)16}$ w. $2\overline{)12}$ C. $3\overline{)24}$

f. $3\overline{)21}$ l. $6\overline{)24}$ r. $9\overline{)27}$ x. $7\overline{)21}$ D. $3\overline{)27}$

Lesson 40

Part 3

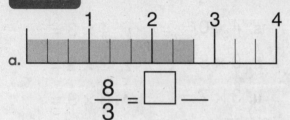

a. $\dfrac{8}{3} = \boxed{} \ \underline{}$

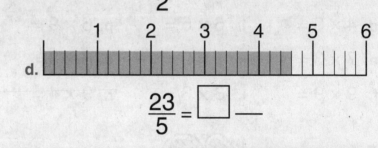

c. $\dfrac{11}{2} = \boxed{} \ \underline{}$

b. $\dfrac{32}{9} = \boxed{} \ \underline{}$

d. $\dfrac{23}{5} = \boxed{} \ \underline{}$

Part 4

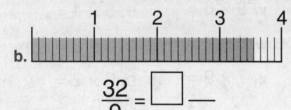

a. $5\overline{\smash{\big)}\,7}$ __

b. $4\overline{\smash{\big)}\,1\,9}$ __

c. $5\overline{\smash{\big)}\,4\,9}$ __

d. $9\overline{\smash{\big)}\,3\,0}$ __

e. $9\overline{\smash{\big)}\,1\,3}$ __

f. $5\overline{\smash{\big)}\,2\,8}$ __

Part 5

a. $3 \times 7 =$
b. $9 \times 9 =$
c. $4 \times 4 =$
d. $6 \times 3 =$
e. $9 \times 7 =$
f. $3 \times 3 =$

g. $4 \times 6 =$
h. $6 \times 9 =$
i. $8 \times 3 =$
j. $6 \times 6 =$
k. $3 \times 4 =$
i. $8 \times 9 =$

m. $4 \times 9 =$
n. $5 \times 7 =$
o. $4 \times 3 =$
p. $1 \times 8 =$
q. $9 \times 6 =$
r. $7 \times 3 =$

s. $0 \times 5 =$
t. $3 \times 8 =$
u. $7 \times 9 =$
v. $6 \times 4 =$
w. $3 \times 9 =$
x. $9 \times 8 =$

y. $5 \times 7 =$
z. $9 \times 4 =$
A. $3 \times 6 =$
B. $8 \times 2 =$
C. $9 \times 3 =$
D. $10 \times 5 =$

Connecting Math Concepts

Lesson 41

Part 1

a. 9 ⟶ 54

d. 9 ⟶ 63

g. 3 ⟶ 21

j. ___ ⟶4 16

b. 3 ⟶7 ___

e. ___ ⟶8 24

h. ___ ⟶4 36

k. 6 ⟶6 ___

c. ___ ⟶6 30

f. 6 ⟶ 24

i. ___ ⟶8 16

l. 7 ⟶5 ___

Part 2

a. 6⟌24

g. 4⟌16

m. 5⟌35

s. 9⟌72

y. 2⟌12

b. 9⟌27

h. 9⟌36

n. 9⟌45

t. 5⟌45

z. 9⟌54

c. 2⟌10

i. 8⟌16

o. 9⟌81

u. 4⟌20

A. 5⟌10

d. 7⟌7

j. 3⟌9

p. 10⟌80

v. 4⟌36

B. 5⟌25

e. 6⟌30

k. 3⟌27

q. 5⟌15

w. 7⟌21

C. 6⟌36

f. 6⟌18

l. 8⟌24

r. 4⟌24

x. 4⟌12

D. 7⟌63

Part 3

a.
$$\begin{array}{r} 3 \\ \times\ 9 \\ \hline \end{array}$$

b.
$$\begin{array}{r} 4 \\ \times\ 2 \\ \hline \end{array}$$

c.
$$\begin{array}{r} 6 \\ \times\ 5 \\ \hline \end{array}$$

d.
$$\begin{array}{r} 7 \\ \times\ 3 \\ \hline \end{array}$$

e.
$$\begin{array}{r} 3 \\ \times\ 90 \\ \hline \end{array}$$

f.
$$\begin{array}{r} 4 \\ \times\ 20 \\ \hline \end{array}$$

g.
$$\begin{array}{r} 6 \\ \times\ 50 \\ \hline \end{array}$$

h.
$$\begin{array}{r} 7 \\ \times\ 30 \\ \hline \end{array}$$

Lesson 41

Part 4

a. $5 \overline{)34} \quad \overset{4}{}$
30

c. $4 \overline{)22} \quad \overset{2}{}$
20

e. $9 \overline{)38} \quad \overset{2}{}$
36

b. $9 \overline{)48} \quad \overset{3}{}$
45

d. $5 \overline{)9} \quad \overset{4}{}$
5

f. $4 \overline{)15} \quad \overset{3}{}$
12

Part 5

a. $3 \times 3 =$

b. $8 \times 9 =$

c. $3 \times 8 =$

d. $2 \times 9 =$

e. $6 \times 10 =$

f. $9 \times 9 =$

g. $4 \times 6 =$

h. $5 \times 4 =$

i. $6 \times 9 =$

j. $7 \times 3 =$

k. $0 \times 4 =$

l. $6 \times 6 =$

m. $1 \times 3 =$

n. $5 \times 5 =$

o. $7 \times 9 =$

p. $5 \times 8 =$

q. $2 \times 4 =$

r. $6 \times 4 =$

s. $2 \times 5 =$

t. $3 \times 4 =$

u. $3 \times 7 =$

v. $3 \times 9 =$

w. $9 \times 5 =$

x. $3 \times 6 =$

y. $1 \times 9 =$

z. $10 \times 0 =$

A. $8 \times 3 =$

B. $9 \times 4 =$

C. $6 \times 3 =$

D. $4 \times 4 =$

Part 6

	Bottles	Cans	Total containers
Food Mart		497	
Grocery Land	353		804
Total for both stores	705		

Lesson 42

Part 1

a. 3 →8→ ___   d. ___ →4→ 12   g. 9 →4→ ___   j. 3 →→ 18

b. ___ →6→ 36   e. 9 →→ 81   h. ___ →8→ 24   k. 2 →9→ ___

c. 9 →6→ ___   f. 3 →7→ ___   i. ___ →8→ 72   l. ___ →7→ 63

Part 2

a. 8 × 3 =   g. 5 × 5 =   m. 4√16   s. 7√49   y. 5√10

b. 9 × 4 =   h. 3√27   n. 9 × 6 =   t. 2√14   z. 7 × 9 =

c. 9√18   i. 8√64   o. 6√36   u. 9√90   A. 3√27

d. 3√21   j. 8 × 9 =   p. 8√72   v. 9 × 9 =   B. 9 × 5 =

e. 4 × 4 =   k. 3 × 7 =   q. 3 × 6 =   w. 5 × 8 =   C. 9√81

f. 3√9   l. 10√100   r. 10√80   x. 9√54   D. 5√45

Part 3

a. 7   b. 2   c. 7   d. 5
   × 5   × 8   × 9   × 6

e. 7   f. 2   g. 7   h. 5
  × 5 0   × 8 0   × 9 0   × 6 0

Lesson 42

Part 4

a. $9\overline{)37}$ b. $2\overline{)11}$ c. $4\overline{)15}$ d. $3\overline{)22}$

Part 5

a. $8 \times 9 =$ g. $9 \times 4 =$ m. $3 \times 3 =$ s. $9 \times 10 =$ y. $9 \times 7 =$

b. $3 \times 7 =$ h. $8 \times 0 =$ n. $5 \times 7 =$ t. $3 \times 9 =$ z. $8 \times 5 =$

c. $4 \times 6 =$ i. $6 \times 4 =$ o. $3 \times 8 =$ u. $7 \times 3 =$ A. $4 \times 4 =$

d. $7 \times 9 =$ j. $9 \times 9 =$ p. $4 \times 9 =$ v. $9 \times 2 =$ B. $6 \times 9 =$

e. $8 \times 3 =$ k. $4 \times 3 =$ q. $6 \times 3 =$ w. $10 \times 10 =$ C. $9 \times 8 =$

f. $7 \times 2 =$ l. $6 \times 6 =$ r. $7 \times 10 =$ x. $1 \times 6 =$ D. $3 \times 6 =$

Lesson 43

Part 1

a. $4\overline{)24}$ g. $8\overline{)16}$ m. $9\overline{)72}$ s. $9\overline{)90}$ y. $3\overline{)6}$

b. $3\overline{)24}$ h. $4\overline{)16}$ n. $9 \times 6 =$ t. $3\overline{)12}$ z. $7\overline{)63}$

c. $8 \times 9 =$ i. $7 \times 3 =$ o. $9\overline{)45}$ u. $9 \times 4 =$ A. $6\overline{)24}$

d. $2\overline{)18}$ j. $7 \times 9 =$ p. $3 \times 9 =$ v. $10\overline{)100}$ B. $5 \times 9 =$

e. $3\overline{)18}$ k. $4\overline{)36}$ q. $3\overline{)9}$ w. $9\overline{)54}$ C. $9\overline{)81}$

f. $4 \times 3 =$ l. $6\overline{)36}$ r. $3\overline{)21}$ x. $3 \times 6 =$ D. $3 \times 8 =$

Lesson 43

Part 2

Fraction	Division
a. $\dfrac{14}{2}$	
b. $\dfrac{18}{9}$	
c. $\dfrac{35}{5}$	
d. $\dfrac{12}{4}$	

Part 3

a. $1 + \dfrac{2}{5}$

b. $4 - \dfrac{3}{2}$

c. $\dfrac{10}{3} + 2$

Part 4

a. $9 \times 5 =$ g. $4 \times 6 =$ m. $9 \times 9 =$ s. $6 \times 4 =$ y. $5 \times 4 =$

b. $3 \times 9 =$ h. $8 \times 3 =$ n. $4 \times 4 =$ t. $0 \times 9 =$ z. $5 \times 8 =$

c. $2 \times 3 =$ i. $7 \times 3 =$ o. $7 \times 9 =$ u. $9 \times 6 =$ A. $3 \times 8 =$

d. $0 \times 4 =$ j. $9 \times 4 =$ p. $10 \times 10 =$ v. $3 \times 10 =$ B. $4 \times 10 =$

e. $3 \times 4 =$ k. $6 \times 3 =$ q. $9 \times 8 =$ w. $1 \times 2 =$ C. $5 \times 1 =$

f. $5 \times 5 =$ l. $2 \times 9 =$ r. $3 \times 7 =$ x. $3 \times 3 =$ D. $8 \times 9 =$

Part 5

	Squirrels	Rabbits	Total of both animals
City Park	146		
Denton Park		137	272
Total for both parks		210	

This table shows the squirrels and rabbits in City Park and Denton Park.

Connecting Math Concepts   Lesson 43   **51**

Lesson

Part 1

a. $8 \times 9 =$

b. $4\overline{)16}$

c. $9 \times 9 =$

d. $3\overline{)21}$

e. $6 \times 6 =$

f. $8\overline{)24}$

g. $6\overline{)24}$

h. $3 \times 9 =$

i. $8 \times 3 =$

j. $9\overline{)72}$

k. $3 \times 6 =$

l. $9\overline{)18}$

m. $9\overline{)54}$

n. $3 \times 8 =$

o. $5\overline{)45}$

p. $4 \times 5 =$

q. $5\overline{)25}$

r. $3 \times 7 =$

s. $9\overline{)63}$

t. $6\overline{)36}$

u. $4 \times 6 =$

v. $3\overline{)24}$

w. $6 \times 4 =$

x. $5\overline{)10}$

y. $4\overline{)12}$

z. $4 \times 9 =$

A. $6\overline{)18}$

B. $9\overline{)36}$

C. $9 \times 7 =$

D. $7\overline{)21}$

Part 2

a. $\dfrac{19}{2} - 6$

b. $5 + \dfrac{3}{7}$

c. $10 - \dfrac{23}{3}$

d. $\dfrac{10}{9} + 4$

Part 3

a. $8 \times 3 =$

b. $7 \times 9 =$

c. $3 \times 4 =$

d. $6 \times 2 =$

e. $6 \times 6 =$

f. $8 \times 5 =$

g. $5 \times 5 =$

h. $6 \times 9 =$

i. $9 \times 5 =$

j. $3 \times 7 =$

k. $5 \times 4 =$

l. $3 \times 3 =$

m. $6 \times 5 =$

n. $2 \times 5 =$

o. $3 \times 6 =$

p. $9 \times 9 =$

q. $9 \times 4 =$

r. $3 \times 8 =$

s. $5 \times 3 =$

t. $4 \times 6 =$

u. $5 \times 7 =$

v. $4 \times 4 =$

w. $7 \times 3 =$

x. $9 \times 6 =$

y. $2 \times 8 =$

z. $10 \times 10 =$

A. $0 \times 9 =$

B. $5 \times 1 =$

C. $6 \times 3 =$

D. $10 \times 7 =$

Lesson 44

	Men	Women	Total adults
The Sloan Building	68	133	
Barlow Offices		95	
Total for both buildings			420

Lesson 45

Part 1

a. $2\overline{)16}$ g. $2\overline{)6}$ m. $9\overline{)36}$ s. $6 \times 3 =$ y. $6\overline{)36}$

b. $4\overline{)16}$ h. $7\overline{)21}$ n. $9 \times 7 =$ t. $6\overline{)24}$ z. $4 \times 6 =$

c. $6\overline{)54}$ i. $4 \times 10 =$ o. $9\overline{)63}$ u. $5\overline{)45}$ A. $9\overline{)27}$

d. $7 \times 2 =$ j. $9\overline{)72}$ p. $7 \times 3 =$ v. $4 \times 4 =$ B. $5\overline{)45}$

e. $3 \times 8 =$ k. $6 \times 6 =$ q. $4\overline{)12}$ w. $9\overline{)81}$ C. $3 \times 7 =$

f. $3\overline{)9}$ l. $5\overline{)35}$ r. $3\overline{)24}$ x. $6\overline{)18}$ D. $10\overline{)80}$

Lesson 45

Part 2

a. $\begin{array}{r} 63 \\ \times\ 30 \\ \hline \end{array}$

b. $\begin{array}{r} 86 \\ \times\ 50 \\ \hline \end{array}$

c. $\begin{array}{r} 39 \\ \times\ 40 \\ \hline \end{array}$

d. $\begin{array}{r} 65 \\ \times\ 60 \\ \hline \end{array}$

Part 3

Fractions	Decimals
a.	.09
b. $\dfrac{5}{10}$	
c. $\dfrac{4}{100}$	
d.	.03
e.	.00
f. $\dfrac{14}{100}$	

Part 4

a. $5 \times 8 =$

b. $3 \times 3 =$

c. $6 \times 2 =$

d. $7 \times 3 =$

e. $10 \times 8 =$

f. $4 \times 4 =$

g. $10 \times 10 =$

h. $8 \times 9 =$

i. $3 \times 8 =$

j. $10 \times 6 =$

k. $6 \times 4 =$

l. $9 \times 3 =$

m. $6 \times 6 =$

n. $8 \times 3 =$

o. $7 \times 2 =$

p. $5 \times 5 =$

q. $4 \times 6 =$

r. $5 \times 3 =$

s. $3 \times 7 =$

t. $9 \times 9 =$

u. $6 \times 9 =$

v. $6 \times 3 =$

w. $9 \times 2 =$

x. $9 \times 8 =$

y. $8 \times 0 =$

z. $4 \times 9 =$

A. $3 \times 6 =$

B. $3 \times 4 =$

C. $5 \times 9 =$

D. $7 \times 9 =$

Lesson 46

Part 1

a. $4 \overline{)}\ 32$

b. $4 \overset{7}{\overline{)}}\ \underline{}$

c. $3 \overline{)}\ 24$

d. $9 \overset{6}{\overline{)}}\ \underline{}$

e. $4 \overline{)}\ 24$

f. $3 \overset{7}{\overline{)}}\ \underline{}$

g. $\underline{} \overset{4}{\overline{)}}\ 36$

h. $9 \overset{8}{\overline{)}}\ \underline{}$

i. $3 \overline{)}\ 18$

j. $4 \overset{8}{\overline{)}}\ \underline{}$

k. $\underline{} \overset{4}{\overline{)}}\ 12$

l. $4 \overline{)}\ 28$

Lesson 46

Part 2

a. $9\overline{)81}$	g. $8\overline{)32}$	m. $6\overline{)24}$	s. $2\overline{)0}$	y. $5\overline{)5}$
b. $6\overline{)36}$	h. $8\overline{)72}$	n. $3\overline{)9}$	t. $6\overline{)6}$	z. $4\overline{)32}$
c. $9\overline{)36}$	i. $9\overline{)54}$	o. $6\overline{)18}$	u. $3\overline{)21}$	A. $9\overline{)54}$
d. $9\overline{)72}$	j. $7\overline{)28}$	p. $9\overline{)27}$	v. $5\overline{)20}$	B. $3\overline{)24}$
e. $6\overline{)6}$	k. $3\overline{)12}$	q. $10\overline{)80}$	w. $4\overline{)28}$	C. $4\overline{)24}$
f. $9\overline{)63}$	l. $5\overline{)35}$	r. $2\overline{)18}$	x. $5\overline{)30}$	D. $3\overline{)18}$

Part 3

Fractions	Decimals
a.	8.09
b. $\dfrac{275}{10}$	
c. $\dfrac{614}{100}$	
d.	70.3
e.	42.50
f. $\dfrac{38}{10}$	

Part 4

a. $8 \times 4 =$	k. $6 \times 5 =$	u. $8 \times 2 =$
b. $2 \times 2 =$	l. $8 \times 5 =$	v. $9 \times 6 =$
c. $4 \times 3 =$	m. $7 \times 10 =$	w. $4 \times 7 =$
d. $6 \times 4 =$	n. $9 \times 9 =$	x. $4 \times 4 =$
e. $5 \times 9 =$	o. $4 \times 6 =$	y. $6 \times 3 =$
f. $3 \times 7 =$	p. $9 \times 4 =$	z. $9 \times 7 =$
g. $2 \times 6 =$	q. $5 \times 5 =$	A. $4 \times 5 =$
h. $8 \times 3 =$	r. $6 \times 6 =$	B. $8 \times 10 =$
i. $7 \times 4 =$	s. $5 \times 7 =$	C. $4 \times 9 =$
j. $2 \times 7 =$	t. $4 \times 8 =$	D. $10 \times 10 =$

Lesson 47

a. $4 \longrightarrow 24$ d. $6 \longrightarrow 24$ g. $\underline{\quad} \overset{7}{\longrightarrow} 28$ j. $9 \longrightarrow 81$

b. $4 \overset{7}{\longrightarrow} \underline{\quad}$ e. $\underline{\quad} \overset{8}{\longrightarrow} 72$ h. $4 \overset{4}{\longrightarrow} \underline{\quad}$ k. $4 \longrightarrow 32$

c. $\underline{\quad} \overset{8}{\longrightarrow} 32$ f. $9 \longrightarrow 63$ i. $\underline{\quad} \overset{6}{\longrightarrow} 24$ l. $\underline{\quad} \overset{3}{\longrightarrow} 18$

Part 2

a. $6\overline{)36}$ g. $9\overline{)54}$ m. $4\overline{)32}$ s. $9\overline{)72}$ y. $5\overline{)45}$

b. $4\overline{)28}$ h. $8\overline{)32}$ n. $5\overline{)25}$ t. $2\overline{)10}$ z. $6\overline{)18}$

c. $3\overline{)12}$ i. $7\overline{)21}$ o. $9\overline{)45}$ u. $5\overline{)45}$ A. $4\overline{)12}$

d. $9\overline{)36}$ j. $6\overline{)24}$ p. $5\overline{)20}$ v. $5\overline{)35}$ B. $3\overline{)9}$

e. $4\overline{)16}$ k. $10\overline{)80}$ q. $7\overline{)28}$ w. $3\overline{)21}$ C. $9\overline{)81}$

f. $9\overline{)63}$ l. $5\overline{)30}$ r. $4\overline{)24}$ x. $9\overline{)72}$ D. $4\overline{)36}$

Lesson 47

Part 3

Fractions	Decimals
a. $\dfrac{3040}{100}$	
b.	30.4
c. $\dfrac{34}{10}$	
d.	5.67
e. $\dfrac{567}{10}$	
f.	567.00

Part 4

a. $4 \times 6 =$

b. $7 \times 0 =$

c. $5 \times 9 =$

d. $8 \times 5 =$

e. $9 \times 4 =$

f. $8 \times 4 =$

g. $5 \times 5 =$

h. $3 \times 7 =$

i. $6 \times 9 =$

j. $3 \times 4 =$

k. $8 \times 3 =$

l. $7 \times 4 =$

m. $9 \times 9 =$

n. $4 \times 3 =$

o. $6 \times 6 =$

p. $6 \times 4 =$

q. $9 \times 7 =$

r. $4 \times 8 =$

s. $3 \times 8 =$

t. $4 \times 7 =$

u. $4 \times 9 =$

v. $9 \times 5 =$

w. $9 \times 8 =$

x. $3 \times 9 =$

y. $7 \times 9 =$

z. $5 \times 7 =$

A. $9 \times 6 =$

B. $7 \times 3 =$

C. $6 \times 3 =$

D. $6 \times 5 =$

Lesson 48

Part 1

a. 4 ⌐→ 32

b. ___ ⌐6→ 24

c. 4 ⌐7→ ___

d. 9 ⌐→ 63

e. ___ ⌐4→ 16

f. 6 ⌐→ 24

g. ___ ⌐8→ 72

h. ___ ⌐7→ 28

i. ___ ⌐9→ 81

j. 4 ⌐→ 24

k. 6 ⌐→ 18

l. 4 ⌐8→ ___

Lesson 48

a. $9\overline{)63}$ g. $5\overline{)35}$ m. $2\overline{)6}$ s. $4\overline{)28}$ y. $5\overline{)5}$

b. $4\overline{)12}$ h. $8\overline{)72}$ n. $6\overline{)24}$ t. $5\overline{)10}$ z. $4\overline{)24}$

c. $6\overline{)36}$ i. $10\overline{)80}$ o. $8\overline{)24}$ u. $4\overline{)32}$ A. $6\overline{)54}$

d. $2\overline{)10}$ j. $3\overline{)9}$ p. $8\overline{)32}$ v. $9\overline{)81}$ B. $2\overline{)16}$

e. $6\overline{)18}$ k. $5\overline{)45}$ q. $9\overline{)36}$ w. $9\overline{)45}$ C. $2\overline{)12}$

f. $7\overline{)63}$ l. $9\overline{)72}$ r. $9\overline{)54}$ x. $5\overline{)30}$ D. $3\overline{)21}$

Part 3

Fractions	Decimals
a. $\dfrac{502}{10}$	
b.	18.6
c. $\dfrac{403}{100}$	
d. $\dfrac{403}{10}$	
e.	4.30
f.	43.0

Part 4

a. $4 \times 8 =$ k. $6 \times 4 =$ u. $3 \times 3 =$

b. $6 \times 6 =$ l. $8 \times 9 =$ v. $8 \times 4 =$

c. $4 \times 9 =$ m. $4 \times 6 =$ w. $7 \times 3 =$

d. $5 \times 9 =$ n. $4 \times 4 =$ x. $9 \times 2 =$

e. $9 \times 7 =$ o. $3 \times 9 =$ y. $4 \times 9 =$

f. $7 \times 4 =$ p. $4 \times 7 =$ z. $5 \times 6 =$

g. $3 \times 7 =$ q. $3 \times 8 =$ A. $4 \times 5 =$

h. $4 \times 3 =$ r. $6 \times 3 =$ B. $8 \times 2 =$

i. $7 \times 5 =$ s. $8 \times 3 =$ C. $5 \times 2 =$

j. $2 \times 10 =$ t. $6 \times 9 =$ D. $9 \times 8 =$

Connecting Math Concepts

Lesson 49

Part 1

a. $4 \overset{\longrightarrow}{} 24$ d. $9 \overset{\longrightarrow}{} 63$ g. $9 \overset{\longrightarrow}{} 36$ j. $4 \overset{8}{\longrightarrow} \underline{}$

b. $\underline{} \overset{7}{\longrightarrow} 28$ e. $\underline{} \overset{8}{\longrightarrow} 72$ h. $7 \overset{\longrightarrow}{} 21$ k. $\underline{} \overset{3}{\longrightarrow} 18$

c. $4 \overset{\longrightarrow}{} 32$ f. $4 \overset{7}{\longrightarrow} \underline{}$ i. $6 \overset{\longrightarrow}{} 24$ l. $8 \overset{3}{\longrightarrow} \underline{}$

Part 2

a. $5 \overline{)5}$ g. $4 \overline{)16}$ m. $3 \overline{)12}$ s. $3 \overline{)27}$ y. $3 \overline{)24}$

b. $2 \overline{)6}$ h. $7 \overline{)63}$ n. $9 \overline{)36}$ t. $9 \overline{)45}$ z. $4 \overline{)24}$

c. $2 \overline{)10}$ i. $6 \overline{)24}$ o. $6 \overline{)36}$ u. $6 \overline{)18}$ A. $6 \overline{)54}$

d. $5 \overline{)35}$ j. $4 \overline{)28}$ p. $2 \overline{)16}$ v. $8 \overline{)32}$ B. $5 \overline{)30}$

e. $9 \overline{)27}$ k. $9 \overline{)72}$ q. $5 \overline{)25}$ w. $9 \overline{)81}$ C. $4 \overline{)12}$

f. $3 \overline{)9}$ l. $7 \overline{)21}$ r. $4 \overline{)32}$ x. $7 \overline{)28}$ D. $3 \overline{)27}$

Part 3

a.
$$\begin{array}{r} 51 \\ \times\ 42 \\ \hline 102 \end{array}$$

b.
$$\begin{array}{r} 38 \\ \times\ 24 \\ \hline 152 \end{array}$$

c.
$$\begin{array}{r} 63 \\ \times\ 52 \\ \hline 126 \end{array}$$

Part 4

a. $3\dfrac{2}{5}$ b. $10\dfrac{7}{8}$ c. $2\dfrac{5}{9}$

Lesson 49

Part 5

a. $2 \times 2 =$ g. $9 \times 6 =$ m. $3 \times 1 =$ s. $3 \times 9 =$ y. $8 \times 4 =$

b. $7 \times 9 =$ h. $7 \times 4 =$ n. $6 \times 3 =$ t. $0 \times 7 =$ z. $9 \times 8 =$

c. $4 \times 8 =$ i. $0 \times 5 =$ o. $3 \times 7 =$ u. $6 \times 4 =$ A. $6 \times 6 =$

d. $5 \times 9 =$ j. $9 \times 9 =$ p. $10 \times 6 =$ v. $4 \times 9 =$ B. $4 \times 4 =$

e. $3 \times 7 =$ k. $5 \times 10 =$ q. $8 \times 3 =$ w. $4 \times 7 =$ C. $7 \times 5 =$

f. $4 \times 3 =$ l. $8 \times 9 =$ r. $9 \times 7 =$ x. $8 \times 1 =$ D. $5 \times 8 =$

Lesson 50

Part 1

a. 9 ⟶ 63   d. 4 ⟶ 24   g. ___ ⟶4 28   j. 9 ⟶ 36

b. 7 ⟶4 ___   e. ___ ⟶9 81   h. 9 ⟶4 ___   k. ___ ⟶8 72

c. ___ ⟶3 12   f. 8 ⟶4 ___   i. 4 ⟶ 16   l. 8 ⟶ 32

Part 2

a. $5\overline{)25}$ g. $6\overline{)24}$ m. $9\overline{)45}$ s. $5\overline{)20}$ y. $4\overline{)16}$

b. $2\overline{)16}$ h. $8\overline{)24}$ n. $6\overline{)18}$ t. $10\overline{)80}$ z. $3\overline{)9}$

c. $5\overline{)45}$ i. $4\overline{)28}$ o. $6\overline{)36}$ u. $3\overline{)18}$ A. $5\overline{)35}$

d. $4\overline{)36}$ j. $8\overline{)32}$ p. $9\overline{)81}$ v. $7\overline{)28}$ B. $3\overline{)21}$

e. $4\overline{)12}$ k. $2\overline{)12}$ q. $4\overline{)32}$ w. $4\overline{)24}$ C. $5\overline{)30}$

f. $3\overline{)21}$ l. $9\overline{)27}$ r. $9\overline{)36}$ x. $3\overline{)24}$ D. $9\overline{)54}$

Lesson 50

Part 3

a.		3 6	b.		7 5	c.		4 9	d.		9 3
	×	5 9		×	2 4		×	6 1		×	4 7
		3 2 4			3 0 0			4 9			6 5 1

Part 4

a. $2 \times 2 =$ g. $7 \times 5 =$ m. $4 \times 6 =$ s. $9 \times 9 =$ y. $8 \times 9 =$

b. $4 \times 8 =$ h. $5 \times 8 =$ n. $8 \times 3 =$ t. $6 \times 6 =$ z. $4 \times 4 =$

c. $6 \times 3 =$ i. $9 \times 7 =$ o. $3 \times 6 =$ u. $9 \times 4 =$ A. $7 \times 9 =$

d. $9 \times 2 =$ j. $3 \times 8 =$ p. $6 \times 9 =$ v. $10 \times 5 =$ B. $7 \times 4 =$

e. $4 \times 7 =$ k. $9 \times 5 =$ q. $3 \times 3 =$ w. $4 \times 0 =$ C. $6 \times 10 =$

f. $3 \times 3 =$ l. $3 \times 7 =$ r. $7 \times 1 =$ x. $9 \times 6 =$ D. $1 \times 5 =$

Lesson 51

Part 1

a. ___ $\overset{4}{\vert}$ 36   d. ___ $\overset{4}{\vert}$ 32   g. ___ $\overset{3}{\vert}$ 24   j. 9 $\vert$ 72

b. 8 $\overset{3}{\vert}$ ___   e. 4 $\vert$ 24   h. 6 $\vert$ 18   k. 4 $\overset{8}{\vert}$ ___

c. 6 $\vert$ 36   f. 7 $\overset{4}{\vert}$ ___   i. ___ $\overset{7}{\vert}$ 28   l. 9 $\vert$ 54

Lesson 51

Part 2

a. $3\overline{)24}$	g. $8\overline{)0}$	m. $4\overline{)12}$	s. $9\overline{)36}$	y. $9\overline{)54}$
b. $2\overline{)10}$	h. $4\overline{)28}$	n. $9\overline{)72}$	t. $4\overline{)16}$	z. $5\overline{)10}$
c. $3\overline{)9}$	i. $5\overline{)20}$	o. $3\overline{)21}$	u. $4\overline{)32}$	A. $9\overline{)81}$
d. $7\overline{)21}$	j. $5\overline{)35}$	p. $6\overline{)24}$	v. $9\overline{)63}$	B. $6\overline{)18}$
e. $7\overline{)28}$	k. $2\overline{)12}$	q. $9\overline{)27}$	w. $3\overline{)12}$	C. $9\overline{)45}$
f. $5\overline{)30}$	l. $7\overline{)7}$	r. $6\overline{)36}$	x. $3\overline{)18}$	D. $8\overline{)32}$

Part 3

a.
$$\begin{array}{r} 35 \\ \times\ 49 \\ \hline 315 \\ \hline \\ \hline \end{array}$$

b.
$$\begin{array}{r} 83 \\ \times\ 92 \\ \hline 166 \\ \hline \\ \hline \end{array}$$

c.
$$\begin{array}{r} 26 \\ \times\ 54 \\ \hline 104 \\ \hline \\ \hline \end{array}$$

Part 4

a. $9 \times 9 =$	g. $7 \times 3 =$	m. $0 \times 4 =$	s. $6 \times 3 =$	y. $9 \times 6 =$
b. $8 \times 3 =$	h. $8 \times 9 =$	n. $4 \times 9 =$	t. $4 \times 6 =$	z. $6 \times 4 =$
c. $7 \times 4 =$	i. $4 \times 8 =$	o. $3 \times 7 =$	u. $4 \times 4 =$	A. $9 \times 8 =$
d. $8 \times 2 =$	j. $6 \times 9 =$	p. $4 \times 7 =$	v. $1 \times 5 =$	B. $3 \times 3 =$
e. $4 \times 9 =$	k. $10 \times 9 =$	q. $9 \times 3 =$	w. $3 \times 6 =$	C. $8 \times 4 =$
f. $6 \times 6 =$	l. $3 \times 4 =$	r. $7 \times 9 =$	x. $4 \times 3 =$	D. $3 \times 8 =$

Lesson 52

Part 1

a. $4\overline{)12}$ g. $3 \times 6 =$ m. $8\overline{)24}$ s. $3\overline{)18}$ y. $6\overline{)36}$

b. $3\overline{)9}$ h. $10\overline{)0}$ n. $1\overline{)5}$ t. $6 \times 6 =$ z. $5\overline{)30}$

c. $4\overline{)28}$ i. $7\overline{)21}$ o. $9\overline{)63}$ u. $9\overline{)72}$ A. $9\overline{)81}$

d. $9 \times 7 =$ j. $9 \times 6 =$ p. $6 \times 4 =$ v. $4 \times 7 =$ B. $5\overline{)35}$

e. $8 \times 4 =$ k. $4\overline{)32}$ q. $3 \times 8 =$ w. $9\overline{)36}$ C. $6\overline{)54}$

f. $9\overline{)54}$ l. $6\overline{)24}$ r. $4\overline{)24}$ x. $2\overline{)10}$ D. $3 \times 7 =$

Part 2

a. $6 \times 2 =$ g. $9 \times 8 =$ m. $4 \times 4 =$ s. $9 \times 9 =$ y. $6 \times 5 =$

b. $4 \times 9 =$ h. $4 \times 7 =$ n. $1 \times 7 =$ t. $3 \times 6 =$ z. $8 \times 4 =$

c. $8 \times 3 =$ i. $9 \times 3 =$ o. $5 \times 5 =$ u. $10 \times 10 =$ A. $3 \times 5 =$

d. $7 \times 2 =$ j. $6 \times 4 =$ p. $7 \times 4 =$ v. $7 \times 3 =$ B. $7 \times 9 =$

e. $6 \times 3 =$ k. $9 \times 5 =$ q. $6 \times 6 =$ w. $4 \times 5 =$ C. $4 \times 3 =$

f. $2 \times 8 =$ l. $4 \times 8 =$ r. $3 \times 3 =$ x. $0 \times 8 =$ D. $9 \times 4 =$

Lesson

Part 1

a. $4\overline{)28}$	g. $8 \times 4 =$	m. $9\overline{)72}$	s. $6 \times 1 =$	y. $9\overline{)63}$
b. $5\overline{)20}$	h. $5\overline{)0}$	n. $3 \times 8 =$	t. $9\overline{)45}$	z. $4\overline{)36}$
c. $9 \times 9 =$	i. $6\overline{)36}$	o. $4\overline{)28}$	u. $9\overline{)54}$	A. $7 \times 7 =$
d. $7 \times 3 =$	j. $4\overline{)12}$	p. $9\overline{)27}$	v. $7 \times 4 =$	B. $6\overline{)18}$
e. $5\overline{)30}$	k. $6\overline{)24}$	q. $6 \times 6 =$	w. $4\overline{)24}$	C. $4\overline{)16}$
f. $8\overline{)24}$	l. $6 \times 3 =$	r. $2\overline{)12}$	x. $4\overline{)32}$	D. $7\overline{)28}$

Part 2

Fraction	Divison
a. $\dfrac{12}{4} =$	$\overline{)}$
b. $\dfrac{24}{6} =$	$\overline{)}$
c.	$9\overline{)18}$
d. $\dfrac{35}{7} =$	$\overline{)}$
e.	$6\overline{)54}$

Part 3

a. $6 \times 4 =$	k. $6 \times 3 =$	u. $9 \times 6 =$
b. $4 \times 3 =$	l. $1 \times 4 =$	v. $5 \times 6 =$
c. $9 \times 4 =$	m. $3 \times 9 =$	w. $2 \times 3 =$
d. $4 \times 4 =$	n. $10 \times 7 =$	x. $9 \times 8 =$
e. $3 \times 8 =$	o. $3 \times 6 =$	y. $3 \times 7 =$
f. $4 \times 7 =$	p. $8 \times 2 =$	z. $7 \times 9 =$
g. $9 \times 2 =$	q. $3 \times 5 =$	A. $6 \times 10 =$
h. $3 \times 3 =$	r. $7 \times 4 =$	B. $8 \times 5 =$
i. $8 \times 4 =$	s. $9 \times 5 =$	C. $6 \times 6 =$
j. $9 \times 7 =$	t. $4 \times 0 =$	D. $6 \times 9 =$

Part 4

a. $5\overline{)235}$	c. $9\overline{)39}$	e. $3\overline{)171}$
b. $3\overline{)95}$	d. $4\overline{)129}$	f. $2\overline{)52}$

Lesson 54

Part 1

a. $4\overline{)32}$ g. $9\overline{)72}$ m. $6 \times 3 =$ s. $3\overline{)9}$ y. $4 \times 3 =$

b. $4\overline{)24}$ h. $3 \times 8 =$ n. $6\overline{)24}$ t. $5\overline{)25}$ z. $9 \times 9 =$

c. $7 \times 4 =$ i. $4\overline{)28}$ o. $4\overline{)12}$ u. $4 \times 4 =$ A. $9 \times 6 =$

d. $9\overline{)54}$ j. $9\overline{)27}$ p. $9\overline{)36}$ v. $5\overline{)30}$ B. $5\overline{)20}$

e. $3\overline{)18}$ k. $6 \times 6 =$ q. $7\overline{)21}$ w. $6\overline{)18}$ C. $6\overline{)36}$

f. $0 \times 5 =$ l. $2\overline{)12}$ r. $8 \times 4 =$ x. $7 \times 3 =$ D. $2\overline{)16}$

Part 2

a. $\dfrac{4}{5} \times \dfrac{3}{9} =$ b. $\dfrac{7}{6} \times \dfrac{5}{6} =$ c. $\dfrac{2}{5} \times \dfrac{4}{1} =$ d. $\dfrac{7}{10} \times \dfrac{9}{8} =$

Part 3

a. $8 \times 4 =$ g. $2 \times 3 =$ m. $4 \times 3 =$ s. $7 \times 9 =$ y. $4 \times 7 =$

b. $9 \times 8 =$ h. $3 \times 4 =$ n. $6 \times 4 =$ t. $8 \times 0 =$ z. $6 \times 10 =$

c. $6 \times 9 =$ i. $6 \times 2 =$ o. $9 \times 6 =$ u. $10 \times 7 =$ A. $8 \times 5 =$

d. $1 \times 4 =$ j. $3 \times 8 =$ p. $3 \times 9 =$ v. $2 \times 4 =$ B. $5 \times 6 =$

e. $7 \times 3 =$ k. $4 \times 4 =$ q. $4 \times 8 =$ w. $7 \times 2 =$ C. $8 \times 2 =$

f. $4 \times 6 =$ l. $7 \times 4 =$ r. $6 \times 3 =$ x. $3 \times 3 =$ D. $9 \times 7 =$

Lesson 55

Part 1

a. $1\overline{)10}$ g. $7\overline{)28}$ m. $7 \times 3 =$ s. $6\overline{)36}$ y. $10 \times 5 =$

b. $6 \times 3 =$ h. $5\overline{)0}$ n. $7 \times 9 =$ t. $8\overline{)8}$ z. $9\overline{)72}$

c. $5\overline{)25}$ i. $6\overline{)18}$ o. $9\overline{)54}$ u. $3 \times 8 =$ A. $4 \times 8 =$

d. $3\overline{)24}$ j. $7 \times 4 =$ p. $4\overline{)28}$ v. $5\overline{)45}$ B. $2\overline{)16}$

e. $4\overline{)24}$ k. $9\overline{)27}$ q. $4\overline{)36}$ w. $6\overline{)24}$ C. $10\overline{)80}$

f. $8 \times 4 =$ l. $4\overline{)12}$ r. $9 \times 9 =$ x. $9\overline{)63}$ D. $4\overline{)32}$

Part 2

a.
$$\begin{array}{r} 62 \\ \times 546 \\ \hline 372 \\ 2480 \\ \hline \end{array}$$

b.
$$\begin{array}{r} 49 \\ \times 326 \\ \hline 294 \\ 980 \\ \hline \end{array}$$

c.
$$\begin{array}{r} 25 \\ \times 428 \\ \hline 200 \\ 500 \\ \hline \end{array}$$

Part 3

a. $\dfrac{7}{8} \times \dfrac{1}{9} =$ b. $\dfrac{10}{7} \times \dfrac{6}{5} =$ c. $\dfrac{2}{3} \times \dfrac{8}{7} =$ d. $\dfrac{5}{2} \times \dfrac{6}{9} =$

Part 4

a. $4 \times 7 =$ g. $9 \times 2 =$ m. $9 \times 4 =$ s. $5 \times 7 =$ y. $7 \times 9 =$

b. $8 \times 3 =$ h. $6 \times 4 =$ n. $4 \times 4 =$ t. $3 \times 9 =$ z. $6 \times 3 =$

c. $4 \times 3 =$ i. $4 \times 8 =$ o. $9 \times 7 =$ u. $4 \times 6 =$ A. $9 \times 8 =$

d. $6 \times 6 =$ j. $9 \times 6 =$ p. $3 \times 6 =$ v. $3 \times 8 =$ B. $8 \times 10 =$

e. $9 \times 5 =$ k. $3 \times 3 =$ q. $7 \times 4 =$ w. $6 \times 9 =$ C. $1 \times 4 =$

f. $3 \times 7 =$ l. $7 \times 2 =$ r. $0 \times 6 =$ x. $8 \times 4 =$ D. $9 \times 3 =$

Lesson 56

Part 1

a. $7 \overline{} \rightarrow 35$ d. $\underline{} \overset{4}{\rightarrow} 32$ g. $\underline{} \overset{5}{\rightarrow} 35$ j. $\underline{} \overset{6}{\rightarrow} 42$

b. $7 \overset{4}{\rightarrow} \underline{}$ e. $7 \overline{} \rightarrow 42$ h. $\underline{} \overset{4}{\rightarrow} 36$ k. $9 \overset{8}{\rightarrow} \underline{}$

c. $7 \overset{6}{\rightarrow} \underline{}$ f. $7 \overline{} \rightarrow 49$ i. $\underline{} \overset{7}{\rightarrow} 49$ l. $\underline{} \overset{7}{\rightarrow} 63$

Part 2

a. $7 \overline{)28}$ g. $4 \overline{)12}$ m. $6 \overline{)24}$ s. $9 \overline{)63}$ y. $4 \overline{)24}$

b. $9 \overline{)45}$ h. $8 \overline{)32}$ n. $8 \overline{)24}$ t. $6 \overline{)36}$ z. $5 \overline{)25}$

c. $5 \overline{)35}$ i. $4 \overline{)16}$ o. $2 \overline{)10}$ u. $6 \overline{)42}$ A. $4 \overline{)28}$

d. $3 \overline{)9}$ j. $7 \overline{)49}$ p. $4 \overline{)36}$ v. $5 \overline{)10}$ B. $5 \overline{)30}$

e. $9 \overline{)36}$ k. $6 \overline{)18}$ q. $3 \overline{)21}$ w. $4 \overline{)32}$ C. $9 \overline{)81}$

f. $7 \overline{)42}$ l. $9 \overline{)27}$ r. $10 \overline{)60}$ x. $9 \overline{)72}$ D. $5 \overline{)5}$

Part 3

a.
```
    9 4
  × 6 1 7
    6 5 8
  9 4 0
```

b.
```
    3 8
  × 5 2 9
    3 4 2
  7 6 0
```

c.
```
    3 4
  × 6 7 8
    2 7 2
  2 3 8 0
```

Copyright © The McGraw-Hill Companies, Inc.

Connecting Math Concepts

Lesson 56   **67**

Lesson 56

Part 4

a. $2 \times 2 =$ g. $8 \times 5 =$ m. $8 \times 3 =$ s. $4 \times 3 =$ y. $6 \times 9 =$

b. $3 \times 6 =$ h. $4 \times 7 =$ n. $5 \times 5 =$ t. $4 \times 6 =$ z. $3 \times 8 =$

c. $5 \times 6 =$ i. $8 \times 9 =$ o. $9 \times 6 =$ u. $9 \times 8 =$ A. $6 \times 3 =$

d. $7 \times 6 =$ j. $4 \times 8 =$ p. $8 \times 4 =$ v. $6 \times 7 =$ B. $9 \times 9 =$

e. $9 \times 5 =$ k. $7 \times 7 =$ q. $3 \times 7 =$ w. $7 \times 9 =$ C. $7 \times 3 =$

f. $6 \times 4 =$ l. $3 \times 3 =$ r. $7 \times 4 =$ x. $6 \times 6 =$ D. $4 \times 4 =$

Lesson 57

Part 1

a. $7 \longrightarrow 42$ c. $9 \xrightarrow{8} \_\_$ e. $\_\_ \xrightarrow{4} 32$ g. $4 \xrightarrow{6} \_\_$

b. $7 \xrightarrow{4} \_\_$ d. $7 \longrightarrow 49$ f. $\_\_ \xrightarrow{6} 42$ h. $7 \xrightarrow{7} \_\_$

Part 2

a. $6\overline{)36}$ g. $3\overline{)21}$ m. $4\overline{)24}$ s. $3\overline{)9}$ y. $6\overline{)6}$

b. $9\overline{)36}$ h. $7\overline{)49}$ n. $4\overline{)32}$ t. $4\overline{)12}$ z. $2\overline{)16}$

c. $7\overline{)28}$ i. $9\overline{)45}$ o. $9\overline{)63}$ u. $4\overline{)16}$ A. $6\overline{)24}$

d. $7\overline{)42}$ j. $2\overline{)10}$ p. $7\overline{)0}$ v. $6\overline{)42}$ B. $5\overline{)20}$

e. $9\overline{)27}$ k. $4\overline{)28}$ q. $9\overline{)72}$ w. $6\overline{)18}$ C. $5\overline{)25}$

f. $8\overline{)32}$ l. $8\overline{)24}$ r. $4\overline{)28}$ x. $5\overline{)45}$ D. $5\overline{)30}$

Lesson 57

Part 3

a.
```
      8 6
  × 5 6 3
      2 5 8
    5 1 6 0
```

b.
```
      1 7
  × 9 7 8
      1 3 6
    1 1 9 0
```

c.
```
      6 3
  × 4 5 9
      5 6 7
    3 1 5 0
```

Part 4

a. 6 × 4 =   g. 7 × 3 =   m. 9 × 8 =   s. 4 × 9 =   y. 6 × 3 =

b. 7 × 7 =   h. 10 × 10 =   n. 3 × 6 =   t. 3 × 4 =   z. 4 × 4 =

c. 4 × 5 =   i. 6 × 7 =   o. 9 × 9 =   u. 7 × 6 =   A. 6 × 9 =

d. 3 × 8 =   j. 9 × 4 =   p. 7 × 9 =   v. 3 × 9 =   B. 3 × 3 =

e. 2 × 0 =   k. 6 × 6 =   q. 1 × 8 =   w. 8 × 3 =   C. 7 × 4 =

f. 4 × 8 =   l. 4 × 7 =   r. 8 × 9 =   x. 9 × 10 =   D. 3 × 7 =

Lesson 58

Part 1

a. 7 ⌐⟶ 28   d. 7 ⌐⟶ 21   g. ___ ⌐⁶⟶ 36   j. ___ ⌐⁴⟶ 28

b. 3 ⌐⁶⟶ ___   e. ___ ⌐³⟶ 12   h. 7 ⌐⟶ 49   k. 7 ⌐⁶⟶ ___

c. ___ ⌐⁶⟶ 54 f. ___ ⌐⁶⟶ 42   i. 8 ⌐⟶ 32   l. 8 ⌐⟶ 24

Lesson 58

Part 2

a. $5\overline{)45}$ g. $8\overline{)32}$ m. $9\overline{)81}$ s. $3\overline{)24}$ y. $2\overline{)16}$

b. $9\overline{)36}$ h. $7\overline{)42}$ n. $4\overline{)32}$ t. $9\overline{)63}$ z. $6\overline{)18}$

c. $4\overline{)12}$ i. $9\overline{)27}$ o. $8\overline{)80}$ u. $4\overline{)16}$ A. $5\overline{)10}$

d. $6\overline{)24}$ j. $3\overline{)21}$ p. $7\overline{)49}$ v. $3\overline{)9}$ B. $4\overline{)24}$

e. $9\overline{)72}$ k. $6\overline{)18}$ q. $3\overline{)12}$ w. $1\overline{)5}$ C. $8\overline{)24}$

f. $4\overline{)28}$ l. $6\overline{)36}$ r. $7\overline{)28}$ x. $2\overline{)0}$ D. $2\overline{)6}$

Part 3

a.
$$\begin{array}{r} 496 \\ \times\ \ 53 \\ \hline \\ \hline \\ \hline \end{array}$$

b.
$$\begin{array}{r} 29 \\ \times\ 641 \\ \hline \\ \hline \\ \hline \\ \hline \end{array}$$

Part 4

a. $10 \times 10 =$ h. $4 \times 8 =$ o. $9 \times 9 =$ v. $4 \times 7 =$

b. $3 \times 8 =$ i. $9 \times 6 =$ p. $3 \times 6 =$ w. $9 \times 8 =$

c. $7 \times 7 =$ j. $8 \times 9 =$ q. $0 \times 5 =$ x. $6 \times 7 =$

d. $2 \times 2 =$ k. $6 \times 6 =$ r. $8 \times 5 =$ y. $5 \times 6 =$

e. $1 \times 8 =$ l. $3 \times 4 =$ s. $9 \times 4 =$ z. $7 \times 9 =$

f. $7 \times 6 =$ m. $3 \times 7 =$ t. $4 \times 6 =$ A. $7 \times 4 =$

g. $5 \times 5 =$ n. $8 \times 4 =$ u. $6 \times 3 =$ B. $9 \times 5 =$

Connecting Math Concepts

Lesson 59

Part 1

a. $8 \xrightarrow{\hspace{1cm}} 24$ d. $\underline{\hspace{0.5cm}} \xrightarrow{4} 32$ g. $9 \xrightarrow{4} \underline{\hspace{0.5cm}}$ j. $\underline{\hspace{0.5cm}} \xrightarrow{7} 42$

b. $\underline{\hspace{0.5cm}} \xrightarrow{4} 28$ e. $7 \xrightarrow{6} \underline{\hspace{0.5cm}}$ h. $\underline{\hspace{0.5cm}} \xrightarrow{3} 18$ k. $7 \xrightarrow{3} \underline{\hspace{0.5cm}}$

c. $\underline{\hspace{0.5cm}} \xrightarrow{3} 21$ f. $4 \xrightarrow{\hspace{1cm}} 24$ i. $\underline{\hspace{0.5cm}} \xrightarrow{7} 49$ l. $6 \xrightarrow{\hspace{1cm}} 36$

Part 2

a. $4\overline{)16}$ g. $9\overline{)63}$ m. $6\overline{)36}$ s. $4\overline{)36}$ y. $6\overline{)42}$

b. $9\overline{)72}$ h. $4\overline{)28}$ n. $6\overline{)24}$ t. $9\overline{)45}$ z. $5\overline{)35}$

c. $4\overline{)24}$ i. $9\overline{)36}$ o. $3\overline{)9}$ u. $3\overline{)12}$ A. $5\overline{)0}$

d. $8\overline{)32}$ j. $7\overline{)49}$ p. $4\overline{)12}$ v. $10\overline{)80}$ B. $2\overline{)16}$

e. $7\overline{)42}$ k. $8\overline{)24}$ q. $6\overline{)18}$ w. $3\overline{)9}$ C. $3\overline{)24}$

f. $5\overline{)5}$ l. $2\overline{)6}$ r. $2\overline{)10}$ x. $7\overline{)21}$ D. $5\overline{)10}$

Part 3

a. $\dfrac{5}{4} - \dfrac{2}{4} =$ c. $\dfrac{8}{3} \times \dfrac{1}{3} =$ e. $\dfrac{10}{9} - \dfrac{8}{9} =$

b. $\dfrac{4}{5} \times \dfrac{2}{5} =$ d. $\dfrac{8}{3} + \dfrac{1}{3} =$ f. $\dfrac{10}{9} \times \dfrac{8}{9} =$

Lesson 59

Part 4

a. $7 \times 7 =$ g. $3 \times 5 =$ m. $4 \times 9 =$ s. $6 \times 9 =$ y. $2 \times 8 =$

b. $3 \times 4 =$ h. $8 \times 4 =$ n. $6 \times 3 =$ t. $4 \times 6 =$ z. $7 \times 9 =$

c. $9 \times 5 =$ i. $3 \times 7 =$ o. $3 \times 3 =$ u. $6 \times 5 =$ A. $3 \times 6 =$

d. $7 \times 6 =$ j. $1 \times 0 =$ p. $9 \times 7 =$ v. $9 \times 4 =$ B. $9 \times 8 =$

e. $3 \times 8 =$ k. $10 \times 10 =$ q. $4 \times 8 =$ w. $8 \times 3 =$ C. $9 \times 9 =$

f. $6 \times 6 =$ l. $7 \times 4 =$ r. $8 \times 9 =$ x. $8 \times 5 =$ D. $7 \times 10 =$

Lesson 60

Part 1

a. ___ →4→ 12   c. 7 →7→ ___   e. 3 →8→ ___   g. 6 →→ 18

b. ___ →3→ 21   d. 6 →7→ ___   f. ___ →7→ 42   h. 4 →7→ ___

Part 2

a. $8\overline{)72}$ g. $4\overline{)16}$ m. $4\overline{)28}$ s. $7\overline{)21}$ y. $1\overline{)10}$

b. $3\overline{)12}$ h. $2\overline{)16}$ n. $6\overline{)54}$ t. $4\overline{)36}$ z. $5\overline{)30}$

c. $7\overline{)42}$ i. $8\overline{)8}$ o. $10\overline{)60}$ u. $6\overline{)36}$ A. $2\overline{)6}$

d. $8\overline{)32}$ j. $6\overline{)42}$ p. $4\overline{)32}$ v. $3\overline{)24}$ B. $7\overline{)49}$

e. $6\overline{)0}$ k. $3\overline{)18}$ q. $5\overline{)35}$ w. $4\overline{)24}$ C. $7\overline{)63}$

f. $3\overline{)21}$ l. $2\overline{)18}$ r. $3\overline{)9}$ x. $2\overline{)14}$ D. $9\overline{)27}$

Lesson 60

Part 3

a. $\dfrac{5}{3} \times \dfrac{5}{2} =$

b. $\dfrac{3}{5} + \dfrac{2}{5} =$

c. $\dfrac{9}{7} - \dfrac{6}{7} =$

d. $\dfrac{4}{3} \times \dfrac{2}{3} =$

e. $\dfrac{3}{4} \times \dfrac{2}{3} =$

f. $\dfrac{4}{3} + \dfrac{2}{3} =$

Part 4

a. $10 \times 10 =$

b. $6 \times 7 =$

c. $9 \times 9 =$

d. $10 \times 3 =$

e. $4 \times 8 =$

f. $7 \times 7 =$

g. $5 \times 5 =$

h. $8 \times 0 =$

i. $2 \times 9 =$

j. $1 \times 8 =$

k. $9 \times 5 =$

l. $7 \times 4 =$

m. $8 \times 3 =$

n. $7 \times 5 =$

o. $6 \times 6 =$

p. $3 \times 2 =$

q. $7 \times 9 =$

r. $2 \times 7 =$

s. $4 \times 7 =$

t. $2 \times 8 =$

u. $5 \times 4 =$

v. $9 \times 10 =$

w. $7 \times 6 =$

x. $8 \times 4 =$

y. $3 \times 6 =$

z. $1 \times 7 =$

A. $8 \times 8 =$

B. $4 \times 6 =$

C. $3 \times 7 =$

D. $6 \times 5 =$

Lesson 61

Part 1

a. $7 \overset{4}{\longrightarrow} \underline{\quad}$

b. $\underline{\quad} \overset{3}{\longrightarrow} 18$

c. $6 \overset{7}{\longrightarrow} \underline{\quad}$

d. $3 \overset{8}{\longrightarrow} \underline{\quad}$

e. $7 \longrightarrow 49$

f. $3 \overset{7}{\longrightarrow} \underline{\quad}$

g. $6 \longrightarrow 36$

h. $\underline{\quad} \overset{9}{\longrightarrow} 36$

Lesson 61

a. 7⟌49　　　g. 8⟌32　　　m. 3⟌24　　　s. 4⟌16　　　y. 2⟌8

b. 2⟌6　　　h. 6⟌18　　　n. 7⟌21　　　t. 9⟌9　　　z. 6⟌42

c. 4⟌12　　　i. 4⟌28　　　o. 9⟌54　　　u. 4⟌24　　　A. 10⟌20

d. 6⟌24　　　j. 6⟌36　　　p. 9⟌81　　　v. 3⟌9　　　B. 2⟌18

e. 4⟌36　　　k. 1⟌5　　　q. 8⟌32　　　w. 4⟌20　　　C. 3⟌15

f. 7⟌42　　　l. 5⟌0　　　r. 5⟌35　　　x. 9⟌63　　　D. 2⟌12

Part 3

	Fraction	Division	Mixed Number
a.		⟌	$9\frac{5}{8}$
b.		⟌	$7\frac{3}{4}$
c.		⟌	$5\frac{4}{7}$

Connecting Math Concepts

Lesson 61

Part 4

a. $6 \times 4 =$

b. $8 \times 3 =$

c. $9 \times 1 =$

d. $7 \times 7 =$

e. $3 \times 9 =$

f. $4 \times 7 =$

g. $3 \times 4 =$

h. $6 \times 7 =$

i. $4 \times 8 =$

j. $2 \times 5 =$

k. $7 \times 3 =$

l. $1 \times 0 =$

m. $6 \times 3 =$

n. $7 \times 4 =$

o. $5 \times 6 =$

p. $4 \times 9 =$

q. $10 \times 7 =$

r. $2 \times 4 =$

s. $9 \times 7 =$

t. $1 \times 6 =$

u. $5 \times 8 =$

v. $6 \times 6 =$

w. $3 \times 8 =$

x. $9 \times 6 =$

y. $10 \times 2 =$

z. $2 \times 2 =$

A. $7 \times 6 =$

B. $1 \times 4 =$

C. $3 \times 3 =$

D. $8 \times 9 =$

Part 5

a.
$$\begin{array}{r} 687 \\ \times\ 551 \\ \hline 34{,}350 \\ + \qquad\quad \\ \hline \end{array}$$

b.
$$\begin{array}{r} 492 \\ \times\ 366 \\ \hline \\ +\,147{,}600 \\ \hline \end{array}$$

Part 6

a. $\dfrac{4}{5} \times \dfrac{3}{5}$

b. $\dfrac{4}{5} - \dfrac{3}{5}$

c. $\dfrac{5}{4} \times \dfrac{5}{3}$

d. $\dfrac{16}{11} + \dfrac{9}{11}$

e. $\dfrac{28}{9} - \dfrac{18}{9}$

f. $\dfrac{1}{8} \times \dfrac{36}{5}$

Lesson

a. $7\overline{)49}$ g. $3\overline{)6}$ m. $3\overline{)9}$ s. $3 \times 7 =$ y. $5\overline{)45}$

b. $3\overline{)0}$ h. $8 \times 3 =$ n. $9\overline{)36}$ t. $2 \times 9 =$ z. $4\overline{)12}$

c. $10\overline{)10}$ i. $6 \times 7 =$ o. $7\overline{)28}$ u. $4 \times 10 =$ A. $6\overline{)24}$

d. $9 \times 9 =$ j. $4\overline{)32}$ p. $6\overline{)42}$ v. $8 \times 2 =$ B. $8\overline{)24}$

e. $4\overline{)16}$ k. $3\overline{)18}$ q. $8\overline{)72}$ w. $6\overline{)54}$ C. $7 \times 6 =$

f. $3\overline{)21}$ l. $7 \times 7 =$ r. $4 \times 6 =$ x. $7\overline{)21}$ D. $7\overline{)42}$

	Fraction	Division	Mixed Number
a.		$\overline{)}$	$4\dfrac{7}{8}$
b.	$\dfrac{8}{6}$	$\overline{)}$	
c.		$\overline{)}$	$3\dfrac{3}{7}$
d.		$9\overline{)24}$	

Lesson 62

Part 3

a. $5 \times 10 =$ g. $0 \times 6 =$ m. $3 \times 6 =$ s. $4 \times 9 =$ y. $7 \times 6 =$

b. $9 \times 7 =$ h. $8 \times 4 =$ n. $2 \times 4 =$ t. $7 \times 2 =$ z. $4 \times 3 =$

c. $3 \times 8 =$ i. $7 \times 3 =$ o. $5 \times 5 =$ u. $3 \times 4 =$ A. $6 \times 4 =$

d. $7 \times 7 =$ j. $9 \times 9 =$ p. $4 \times 6 =$ v. $8 \times 9 =$ B. $9 \times 6 =$

e. $1 \times 9 =$ k. $4 \times 7 =$ q. $9 \times 10 =$ w. $4 \times 8 =$ C. $10 \times 4 =$

f. $6 \times 7 =$ l. $8 \times 3 =$ r. $8 \times 2 =$ x. $9 \times 5 =$ D. $3 \times 9 =$

Part 4

a.
```
      476
  ×   929
  ─────────

  +428,400
  ─────────
```

b.
```
      364
  ×   550
  ─────────
    18,200
  +
  ─────────
```

Part 5

a. $\dfrac{7}{2} + \dfrac{7}{5}$ c. $\dfrac{6}{9} \times \dfrac{5}{6}$ e. $\dfrac{5}{4} - \dfrac{3}{9}$

b. $\dfrac{5}{4} - \dfrac{3}{4}$ d. $\dfrac{2}{7} + \dfrac{5}{7}$ f. $\dfrac{5}{4} \times \dfrac{3}{9}$

Lesson

a. $8\overline{)24}$ g. $6 \times 7 =$ m. $7\overline{)70}$ s. $1 \times 5 =$ y. $7\overline{)14}$

b. $3 \times 7 =$ h. $8 \times 9 =$ n. $1\overline{)8}$ t. $7\overline{)63}$ z. $3 \times 8 =$

c. $6\overline{)42}$ i. $4 \times 6 =$ o. $3\overline{)21}$ u. $5 \times 6 =$ A. $10\overline{)0}$

d. $4\overline{)16}$ j. $4\overline{)32}$ p. $6 \times 6 =$ v. $3\overline{)18}$ B. $4\overline{)36}$

e. $7 \times 7 =$ k. $9\overline{)54}$ q. $7\overline{)42}$ w. $8\overline{)32}$ C. $10 \times 10 =$

f. $6\overline{)18}$ l. $7\overline{)49}$ r. $5 \times 3 =$ x. $9\overline{)81}$ D. $7 \times 6 =$

Part 2

a. $\dfrac{3}{5} - \dfrac{3}{4} =$ c. $\dfrac{3}{8} \times \dfrac{3}{4} =$ e. $\dfrac{13}{5} - \dfrac{5}{8} =$

b. $\dfrac{5}{3} - \dfrac{4}{3} =$ d. $\dfrac{10}{7} + \dfrac{25}{7} =$ f. $\dfrac{2}{9} \times \dfrac{30}{7} =$

Lesson 63

Part 3

	Fraction	Division	Mixed Number
a.		$\overline{}$	$3\frac{5}{9}$
b.		$8\overline{)21}$	
c.	$\frac{39}{7}$	$\overline{}$	
d.		$\overline{}$	$10\frac{2}{5}$

Part 4

a. $4 \times 7 =$	g. $5 \times 1 =$	m. $5 \times 4 =$	s. $4 \times 3 =$	y. $8 \times 9 =$
b. $3 \times 8 =$	h. $7 \times 6 =$	n. $3 \times 3 =$	t. $7 \times 4 =$	z. $4 \times 4 =$
c. $7 \times 7 =$	i. $6 \times 3 =$	o. $8 \times 2 =$	u. $6 \times 5 =$	A. $5 \times 2 =$
d. $9 \times 4 =$	j. $2 \times 9 =$	p. $6 \times 7 =$	v. $2 \times 6 =$	B. $3 \times 7 =$
e. $0 \times 10 =$	k. $6 \times 4 =$	q. $5 \times 9 =$	w. $10 \times 3 =$	C. $6 \times 6 =$
f. $8 \times 4 =$	l. $7 \times 3 =$	r. $8 \times 3 =$	x. $9 \times 8 =$	D. $9 \times 6 =$

Lesson 64

Part 1

a. $9 \times 9 =$ g. $2\overline{)14}$ m. $9 \times 8 =$ s. $7\overline{)49}$ y. $6 \times 3 =$

b. $6\overline{)36}$ h. $4 \times 6 =$ n. $4\overline{)36}$ t. $6 \times 7 =$ z. $2\overline{)18}$

c. $7 \times 7 =$ i. $8\overline{)32}$ o. $5\overline{)15}$ u. $8\overline{)24}$ A. $8\overline{)72}$

d. $3\overline{)24}$ j. $4\overline{)16}$ p. $7 \times 6 =$ v. $2\overline{)0}$ B. $6\overline{)24}$

e. $4\overline{)28}$ k. $5\overline{)5}$ q. $3\overline{)30}$ w. $9 \times 7 =$ C. $3\overline{)18}$

f. $6\overline{)42}$ l. $7 \times 3 =$ r. $9\overline{)63}$ x. $3 \times 8 =$ D. $7\overline{)21}$

Part 2

	Fraction	Division	Mixed Number	
a.		$\overline{)}$	$24\frac{1}{4}$	
b.	$\frac{103}{9}$	$\overline{)}$		
c.		$5\overline{)537}$		
d.		$\overline{)}$	$215\frac{2}{3}$	

Lesson 64

Part 3

a. 7 × 6 =   g. 7 × 7 =   m. 7 × 4 =   s. 4 × 6 =   y. 9 × 1 =

b. 5 × 3 =   h. 4 × 8 =   n. 5 × 6 =   t. 6 × 7 =   z. 4 × 7 =

c. 3 × 8 =   i. 3 × 6 =   o. 9 × 3 =   u. 1 × 6 =   A. 9 × 6 =

d. 4 × 3 =   j. 5 × 2 =   p. 4 × 5 =   v. 8 × 3 =   B. 2 × 5 =

e. 6 × 6 =   k. 6 × 9 =   q. 9 × 7 =   w. 4 × 9 =   C. 3 × 4 =

f. 0 × 8 =   l. 7 × 3 =   r. 10 × 8 =   x. 3 × 3 =   D. 8 × 9 =

Lesson 65

Part 1

a. 4 × 7 =   g. 3 × 8 =   m. $8\overline{)32}$ s. $7\overline{)63}$ y. $2\overline{)16}$

b. $3\overline{)21}$ h. 7 × 6 =   n. $9\overline{)81}$ t. $2\overline{)18}$ z. $8\overline{)80}$

c. $9\overline{)9}$ i. $4\overline{)28}$ o. $3\overline{)15}$ u. 6 × 9 =   A. $5\overline{)45}$

d. 9 × 7 =   j. $7\overline{)49}$ p. 4 × 8 =   v. $7\overline{)28}$ B. 3 × 7 =

e. $6\overline{)24}$ k. $3\overline{)18}$ q. $9\overline{)36}$ w. $3\overline{)27}$ C. 8 × 9 =

f. 7 × 7 =   l. $6\overline{)42}$ r. 6 × 7 =   x. $6\overline{)36}$ D. $6\overline{)0}$

Lesson 65

Part 2

Fraction	Division	Mixed or Whole Number	
a.	⌐	$120\dfrac{5}{6}$	
b.	3⟌651		
c. $\dfrac{916}{3}$	⌐		

Part 3

a. $9 \times 6 =$	g. $4 \times 9 =$	m. $8 \times 4 =$	s. $3 \times 7 =$	y. $8 \times 3 =$
b. $7 \times 3 =$	h. $8 \times 9 =$	n. $2 \times 9 =$	t. $4 \times 3 =$	z. $2 \times 4 =$
c. $8 \times 1 =$	i. $6 \times 3 =$	o. $6 \times 7 =$	u. $0 \times 10 =$	A. $6 \times 9 =$
d. $7 \times 6 =$	j. $2 \times 5 =$	p. $7 \times 5 =$	v. $5 \times 5 =$	B. $4 \times 4 =$
e. $3 \times 3 =$	k. $7 \times 7 =$	q. $6 \times 6 =$	w. $7 \times 9 =$	C. $4 \times 7 =$
f. $4 \times 8 =$	l. $4 \times 6 =$	r. $7 \times 4 =$	x. $10 \times 7 =$	D. $9 \times 5 =$

Connecting Math Concepts

Lesson 66

Part 1

a. $6 \xrightarrow{3} \underline{\quad}$ d. $7 \xrightarrow{} 56$ g. $6 \xrightarrow{} 24$ j. $7 \xrightarrow{} 49$

b. $\underline{\quad} \xrightarrow{8} 64$ e. $6 \xrightarrow{8} \underline{\quad}$ h. $7 \xrightarrow{8} \underline{\quad}$ k. $\underline{\quad} \xrightarrow{5} 20$

c. $3 \xrightarrow{8} \underline{\quad}$ f. $\underline{\quad} \xrightarrow{8} 56$ i. $8 \xrightarrow{} 32$ l. $6 \xrightarrow{7} \underline{\quad}$

Part 2

a. $8\overline{)56}$	g. $6\overline{)48}$	m. $3\overline{)18}$	s. $4\overline{)12}$	y. $9\overline{)45}$
b. $4\overline{)28}$	h. $12\overline{)12}$	n. $7\overline{)49}$	t. $7\overline{)56}$	z. $8\overline{)8}$
c. $6\overline{)30}$	i. $6\overline{)54}$	o. $8\overline{)72}$	u. $4\overline{)24}$	A. $3\overline{)21}$
d. $7\overline{)42}$	j. $9\overline{)36}$	p. $10\overline{)0}$	v. $8\overline{)16}$	B. $5\overline{)15}$
e. $5\overline{)35}$	k. $3\overline{)24}$	q. $7\overline{)21}$	w. $8\overline{)72}$	C. $7\overline{)70}$
f. $1\overline{)9}$	l. $8\overline{)64}$	r. $3\overline{)27}$	x. $8\overline{)32}$	D. $7\overline{)56}$

Part 3

	Fraction	Division	Mixed or Whole Number	
a.	$\dfrac{936}{9}$	$\overline{)}$		
b.		$\overline{)}$	$215\dfrac{1}{4}$	
c.		$3\overline{)422}$		

Lesson 66

Part 4

a. $9 \times 7 =$ g. $6 \times 7 =$ m. $3 \times 8 =$ s. $9 \times 9 =$ y. $8 \times 1 =$

b. $3 \times 4 =$ h. $8 \times 8 =$ n. $4 \times 6 =$ t. $4 \times 7 =$ z. $5 \times 6 =$

c. $2 \times 2 =$ i. $10 \times 10 =$ o. $6 \times 8 =$ u. $8 \times 3 =$ A. $7 \times 3 =$

d. $7 \times 8 =$ j. $3 \times 0 =$ p. $3 \times 7 =$ v. $7 \times 7 =$ B. $6 \times 6 =$

e. $9 \times 6 =$ k. $8 \times 4 =$ q. $8 \times 7 =$ w. $6 \times 3 =$ C. $9 \times 3 =$

f. $8 \times 6 =$ l. $4 \times 9 =$ r. $9 \times 8 =$ x. $10 \times 9 =$ D. $4 \times 8 =$

Lesson 67

Part 1

a. ___ $\overset{8}{\longrightarrow}$ 48   d. ___ $\overset{8}{\longrightarrow}$ 24   g. 6 $\overset{\quad}{\longrightarrow}$ 48   j. ___ $\overset{7}{\longrightarrow}$ 21

b. ___ $\overset{6}{\longrightarrow}$ 24   e. 3 $\overset{9}{\longrightarrow}$ ___   h. 8 $\overset{\quad}{\longrightarrow}$ 64   k. 8 $\overset{4}{\longrightarrow}$ ___

c. 8 $\overset{7}{\longrightarrow}$ ___   f. ___ $\overset{7}{\longrightarrow}$ 42   i. 3 $\overset{6}{\longrightarrow}$ ___   l. ___ $\overset{9}{\longrightarrow}$ 81

Part 2

a. $8\overline{)64}$ g. $6\overline{)24}$ m. $6\overline{)54}$ s. $6\overline{)48}$ y. $6\overline{)18}$

b. $2\overline{)18}$ h. $7\overline{)63}$ n. $4\overline{)0}$ t. $2\overline{)6}$ z. $6\overline{)42}$

c. $7\overline{)56}$ i. $7\overline{)42}$ o. $6\overline{)30}$ u. $4\overline{)32}$ A. $4\overline{)16}$

d. $5\overline{)45}$ j. $4\overline{)28}$ p. $3\overline{)24}$ v. $9\overline{)36}$ B. $8\overline{)24}$

e. $3\overline{)12}$ k. $10\overline{)100}$ q. $5\overline{)25}$ w. $5\overline{)5}$ C. $8\overline{)56}$

f. $7\overline{)49}$ l. $2\overline{)14}$ r. $3\overline{)21}$ x. $8\overline{)72}$ D. $9\overline{)81}$

Connecting Math Concepts

Lesson 67

Part 3

a. 1 × 10 =   g. 8 × 8 =   m. 7 × 7 =   s. 8 × 9 =   y. 8 × 7 =

b. 3 × 8 =   h. 8 × 6 =   n. 6 × 4 =   t. 3 × 7 =   z. 2 × 8 =

c. 2 × 9 =   i. 3 × 4 =   o. 8 × 4 =   u. 4 × 9 =   A. 5 × 3 =

d. 7 × 8 =   j. 9 × 3 =   p. 6 × 9 =   v. 7 × 4 =   B. 4 × 4 =

e. 4 × 5 =   k. 7 × 6 =   q. 4 × 2 =   w. 6 × 2 =   C. 9 × 0 =

f. 7 × 9 =   l. 6 × 3 =   r. 8 × 3 =   x. 6 × 6 =   D. 7 × 5 =

Lesson 68

Part 1

a. 7 ⟶ 49   d. 6 ⟶ 42   g. ___ $\xrightarrow{8}$ 56   j. ___ $\xrightarrow{6}$ 36

b. 4 $\xrightarrow{8}$ ___   e. 8 $\xrightarrow{8}$ ___   h. 3 ⟶ 27   k. ___ $\xrightarrow{7}$ 21

c. 2 ⟶ 18   f. 6 ⟶ 48   i. 7 $\xrightarrow{9}$ ___   l. 3 $\xrightarrow{6}$ ___

Part 2

a. $6\overline{)42}$ g. $6\overline{)24}$ m. $7\overline{)63}$ s. $8\overline{)56}$ y. $5\overline{)10}$

b. $3\overline{)12}$ h. $8\overline{)48}$ n. $1\overline{)0}$ t. $3\overline{)24}$ z. $9\overline{)54}$

c. $7\overline{)35}$ i. $7\overline{)21}$ o. $4\overline{)28}$ u. $9\overline{)36}$ A. $6\overline{)60}$

d. $3\overline{)18}$ j. $9\overline{)81}$ p. $9\overline{)72}$ v. $7\overline{)42}$ B. $4\overline{)24}$

e. $8\overline{)64}$ k. $1\overline{)1}$ q. $6\overline{)36}$ w. $5\overline{)50}$ C. $6\overline{)48}$

f. $7\overline{)56}$ l. $7\overline{)49}$ r. $2\overline{)16}$ x. $8\overline{)32}$ D. $8\overline{)8}$

Lesson 68

Part 3

a.

c.

e.

g.

b.

d.

f.

Part 4

a. $6 \times 8 =$ g. $7 \times 8 =$ m. $3 \times 7 =$ s. $8 \times 7 =$ y. $6 \times 7 =$

b. $3 \times 3 =$ h. $6 \times 3 =$ n. $6 \times 10 =$ t. $5 \times 9 =$ z. $4 \times 8 =$

c. $6 \times 4 =$ i. $8 \times 3 =$ o. $5 \times 3 =$ u. $8 \times 6 =$ A. $7 \times 6 =$

d. $7 \times 5 =$ j. $6 \times 6 =$ p. $4 \times 4 =$ v. $4 \times 9 =$ B. $3 \times 6 =$

e. $8 \times 8 =$ k. $4 \times 7 =$ q. $2 \times 9 =$ w. $0 \times 10 =$ C. $8 \times 9 =$

f. $9 \times 9 =$ l. $9 \times 1 =$ r. $7 \times 9 =$ x. $7 \times 7 =$ D. $2 \times 5 =$

Lesson 69

Part 1

a. ___ ⌐8→ 64   d. 8 ⌐8→ ___   g. 6 ⌐→ 48   j. ___ ⌐7→ 21

b. 7 ⌐9→ ___   e. 7 ⌐→ 56   h. 4 ⌐8→ ___   k. 2 ⌐→ 10

c. 7 ⌐→ 49   f. 4 ⌐→ 24   i. ___ ⌐7→ 42   l. 7 ⌐8→ ___

Connecting Math Concepts

Lesson 69

Part 2

a. 7 ⟌ 28   g. 3 ⟌ 18   m. 7 ⟌ 49   s. 10 ⟌ 0   y. 3 ⟌ 21

b. 4 ⟌ 36   h. 3 ⟌ 24   n. 5 ⟌ 35   t. 6 ⟌ 42   z. 4 ⟌ 16

c. 6 ⟌ 48   i. 2 ⟌ 16   o. 7 ⟌ 56   u. 4 ⟌ 24   A. 8 ⟌ 64

d. 8 ⟌ 72   j. 4 ⟌ 32   p. 5 ⟌ 5   v. 2 ⟌ 14   B. 1 ⟌ 10

e. 3 ⟌ 27   k. 7 ⟌ 63   q. 5 ⟌ 30   w. 9 ⟌ 81   C. 8 ⟌ 16

f. 9 ⟌ 54   l. 9 ⟌ 45   r. 2 ⟌ 18   x. 8 ⟌ 48   D. 6 ⟌ 36

Part 3

a. $8 \times 7 =$ g. $8 \times 3 =$ m. $3 \times 8 =$ s. $6 \times 7 =$ y. $7 \times 9 =$

b. $6 \times 8 =$ h. $8 \times 8 =$ n. $6 \times 3 =$ t. $2 \times 9 =$ z. $2 \times 6 =$

c. $7 \times 5 =$ i. $4 \times 6 =$ o. $5 \times 5 =$ u. $1 \times 1 =$ A. $8 \times 4 =$

d. $3 \times 6 =$ j. $6 \times 9 =$ p. $7 \times 8 =$ v. $9 \times 3 =$ B. $6 \times 10 =$

e. $4 \times 7 =$ k. $7 \times 7 =$ q. $3 \times 4 =$ w. $5 \times 2 =$ C. $3 \times 7 =$

f. $8 \times 9 =$ l. $4 \times 2 =$ r. $0 \times 8 =$ x. $4 \times 8 =$ D. $5 \times 9 =$

Lesson 70

Part 1

a. 3 ⌐6→ ___   d. 6 ⌐8→ ___   g. ___ ⌐6→ 24   j. ___ ⌐8→ 64

b. 7 ⌐⟹→ 56   e. 8 ⌐8→ ___   h. 7 ⌐⟹→ 42   k. 3 ⌐7→ ___

c. ___ ⌐9→ 27   f. 7 ⌐⟹→ 49   i. 4 ⌐7→ ___   l. 8 ⌐⟹→ 56

Lesson 70

Part 2

a. $9\overline{)63}$ g. $3\overline{)24}$ m. $8\overline{)64}$ s. $4\overline{)4}$ y. $8\overline{)32}$

b. $4\overline{)16}$ h. $1\overline{)9}$ n. $7\overline{)28}$ t. $5\overline{)25}$ z. $8\overline{)56}$

c. $6\overline{)42}$ i. $7\overline{)56}$ o. $3\overline{)21}$ u. $9\overline{)36}$ A. $6\overline{)24}$

d. $4\overline{)32}$ j. $9\overline{)54}$ p. $7\overline{)56}$ v. $4\overline{)28}$ B. $3\overline{)12}$

e. $6\overline{)0}$ k. $9\overline{)45}$ q. $4\overline{)24}$ w. $10\overline{)80}$ C. $9\overline{)18}$

f. $7\overline{)49}$ l. $6\overline{)36}$ r. $8\overline{)48}$ x. $6\overline{)18}$ D. $3\overline{)21}$

Part 3

a. $4 \times \underline{\quad} = 24$ c. $3 \times 6 = \underline{\quad}$ e. $6 \times 4 = \underline{\quad}$

b. $\underline{\quad} \times 9 = 45$ d. $\underline{\quad} \times 7 = 14$ f. $3 \times \underline{\quad} = 21$

Part 4

a. $\dfrac{10}{3} \times \dfrac{2}{2} = \dfrac{20}{6}$ c. $\dfrac{3}{2} \times \dfrac{8}{9} = \dfrac{24}{18}$

$\underline{\quad} = \underline{\quad}$ $\underline{\quad} = \underline{\quad}$

b. $\dfrac{9}{4} \times \dfrac{6}{5} = \dfrac{54}{20}$ d. $\dfrac{5}{12} \times \dfrac{11}{11} = \dfrac{55}{132}$

$\underline{\quad} = \underline{\quad}$ $\underline{\quad} = \underline{\quad}$

Connecting Math Concepts

Lesson 70

Part 5

a. $7 \times 9 =$	g. $7 \times 8 =$	m. $6 \times 9 =$	s. $6 \times 7 =$	y. $7 \times 5 =$
b. $8 \times 6 =$	h. $4 \times 9 =$	n. $4 \times 4 =$	t. $6 \times 4 =$	z. $6 \times 8 =$
c. $7 \times 3 =$	i. $7 \times 6 =$	o. $6 \times 10 =$	u. $8 \times 4 =$	A. $7 \times 7 =$
d. $4 \times 6 =$	j. $8 \times 5 =$	p. $4 \times 3 =$	v. $5 \times 7 =$	B. $2 \times 5 =$
e. $2 \times 0 =$	k. $4 \times 7 =$	q. $3 \times 8 =$	w. $3 \times 0 =$	C. $6 \times 3 =$
f. $8 \times 8 =$	l. $3 \times 6 =$	r. $9 \times 8 =$	x. $6 \times 1 =$	D. $9 \times 9 =$

Lesson 71

Part 1

a. $6 \times 8 =$	g. $7 \times 8 =$	m. $4 \times 8 =$	s. $4 \overline{)24}$	y. $5 \overline{)45}$
b. $3 \overline{)21}$	h. $8 \times 8 =$	n. $7 \overline{)56}$	t. $5 \overline{)25}$	z. $9 \overline{)72}$
c. $7 \overline{)42}$	i. $2 \overline{)8}$	o. $3 \times 7 =$	u. $8 \overline{)32}$	A. $7 \times 7 =$
d. $3 \times 8 =$	j. $9 \overline{)81}$	p. $6 \overline{)6}$	v. $8 \overline{)64}$	B. $6 \overline{)18}$
e. $6 \overline{)48}$	k. $7 \overline{)63}$	q. $3 \overline{)0}$	w. $7 \times 4 =$	C. $6 \times 6 =$
f. $6 \times 7 =$	l. $4 \overline{)28}$	r. $3 \overline{)24}$	x. $7 \overline{)49}$	D. $4 \overline{)36}$

Part 2

a. $9 \times \underline{} = 27$	c. $5 \times 6 = \underline{}$	e. $8 \times 2 = \underline{}$
b. $\underline{} \times 6 = 54$	d. $\underline{} \times 3 = 21$	f. $8 \times \underline{} = 32$

Lesson 71

Part 3

a. $8 \times \dfrac{9}{10} = \dfrac{72}{10}$ ___ ___

c. $\dfrac{7}{5} \times \dfrac{15}{12} = \dfrac{105}{60}$ ___ ___

e. $\dfrac{5}{7} \times \dfrac{14}{16} = \dfrac{70}{112}$ ___ ___

b. $\dfrac{3}{4} \times \dfrac{13}{13} = \dfrac{39}{52}$ ___ ___

d. $2 \times \dfrac{27}{27} = \dfrac{54}{27}$ ___ ___

f. $\dfrac{8}{11} \times \dfrac{9}{7} = \dfrac{72}{77}$ ___ ___

Part 4

a. $7 \times 7 =$

b. $8 \times 8 =$

c. $6 \times 9 =$

d. $7 \times 8 =$

e. $4 \times 6 =$

f. $10 \times 10 =$

g. $8 \times 6 =$

h. $4 \times 7 =$

i. $3 \times 6 =$

j. $8 \times 4 =$

k. $3 \times 7 =$

l. $5 \times 2 =$

m. $7 \times 9 =$

n. $0 \times 5 =$

o. $8 \times 7 =$

p. $2 \times 9 =$

q. $8 \times 1 =$

r. $6 \times 5 =$

s. $4 \times 8 =$

t. $2 \times 5 =$

u. $5 \times 7 =$

v. $1 \times 7 =$

w. $4 \times 4 =$

x. $8 \times 3 =$

y. $6 \times 8 =$

z. $9 \times 0 =$

A. $9 \times 9 =$

B. $3 \times 4 =$

C. $5 \times 8 =$

D. $6 \times 7 =$

Lesson 72

Part 1

a. $8 \times 6 =$

b. $3 \times 4 =$

c. $6 \times 7 =$

d. $4 \times 8 =$

e. $5 \times 5 =$

f. $4 \times 7 =$

g. $5 \times 1 =$

h. $8 \times 8 =$

i. $6 \times 9 =$

j. $7 \times 8 =$

k. $6 \times 6 =$

l. $3 \times 2 =$

m. $7 \times 3 =$

n. $5 \times 9 =$

o. $8 \times 9 =$

p. $2 \times 6 =$

q. $8 \times 7 =$

r. $7 \times 6 =$

s. $0 \times 4 =$

t. $3 \times 3 =$

u. $7 \times 7 =$

v. $10 \times 10 =$

w. $6 \times 3 =$

x. $2 \times 0 =$

y. $5 \times 3 =$

z. $6 \times 8 =$

A. $7 \times 4 =$

B. $4 \times 4 =$

C. $2 \times 7 =$

D. $3 \times 8 =$

     Connecting Math Concepts

Lesson 72

Part 2

Independent Work

⌐	÷
a. 3⌐1 2 6	÷
b. 4⌐8 0	÷
c. 9⌐1 2 6 9	÷

Part 3

a. $\dfrac{1}{7} \times \dfrac{6}{6} = \dfrac{6}{42}$ ___ ___

d. $9 \times \dfrac{13}{14} = \dfrac{117}{14}$ ___ ___

b. $4 \times \dfrac{25}{24} = \dfrac{100}{24}$ ___ ___

e. $\dfrac{5}{8} \times \dfrac{51}{49} = \dfrac{255}{392}$ ___ ___

c. $\dfrac{3}{2} \times \dfrac{19}{19} = \dfrac{57}{38}$ ___ ___

Part 4

a. $4 \times 8 =$ g. $48 \div 6 =$ m. $18 \div 9 =$ s. $1 \times 10 =$ y. $27 \div 3 =$

b. $42 \div 6 =$ h. $4 \times 5 =$ n. $28 \div 4 =$ t. $36 \div 4 =$ z. $7 \times 8 =$

c. $18 \div 3 =$ i. $9 \times 3 =$ o. $42 \div 7 =$ u. $35 \div 7 =$ A. $48 \div 8 =$

d. $0 \div 8 =$ j. $64 \div 8 =$ p. $2 \times 9 =$ v. $16 \div 2 =$ B. $4 \times 7 =$

e. $32 \div 8 =$ k. $24 \div 4 =$ q. $56 \div 7 =$ w. $72 \div 8 =$ C. $24 \div 8 =$

f. $7 \times 8 =$ l. $7 \times 7 =$ r. $36 \div 6 =$ x. $0 \times 3 =$ D. $3 \times 7 =$

Lesson 73

Part 1

a. 6⌐→48   d. 6⌐⁷→___   g. 7⌐⁷→49   j. 3⌐⁷→___

b. ___⌐⁷→28   e. ___⌐⁴→32   h. ___⌐⁸→24   k. ___⌐⁹→27

c. 8⌐⁹→___   f. 8⌐→64   i. 7⌐⁸→___   l. 4⌐→36

Lesson 73

Part 2

a. 72 ÷ 8 =	g. 36 ÷ 6 =	m. 12 ÷ 6 =	s. 81 ÷ 9 =	y. 64 ÷ 8 =
b. 3 ÷ 3 =	h. 49 ÷ 7 =	n. 56 ÷ 7 =	t. 24 ÷ 3 =	z. 40 ÷ 5 =
c. 54 ÷ 6 =	i. 24 ÷ 6 =	o. 45 ÷ 5 =	u. 18 ÷ 2 =	A. 16 ÷ 8 =
d. 28 ÷ 4 =	j. 48 ÷ 6 =	p. 0 ÷ 3 =	v. 21 ÷ 7 =	B. 27 ÷ 3 =
e. 12 ÷ 4 =	k. 25 ÷ 5 =	q. 63 ÷ 9 =	w. 9 ÷ 3 =	C. 36 ÷ 9 =
f. 90 ÷ 10 =	l. 42 ÷ 6 =	r. 18 ÷ 3 =	x. 32 ÷ 4 =	D. 48 ÷ 8 =

Part 3

Fraction	⌐	÷	Mixed or Whole Number
a.			$3\frac{5}{9}$
b.		39 ÷ 5	
c. $\frac{20}{4}$			
d.			$9\frac{2}{6}$

Part 4

a. 3 × 5 =	g. 6 × 7 =	m. 9 × 9 =	s. 3 × 7 =	y. 9 × 8 =
b. 4 × 6 =	h. 4 × 8 =	n. 2 × 7 =	t. 5 × 9 =	z. 4 × 4 =
c. 8 × 8 =	i. 2 × 3 =	o. 8 × 7 =	u. 7 × 1 =	A. 6 × 8 =
d. 7 × 4 =	j. 3 × 8 =	p. 3 × 6 =	v. 6 × 3 =	B. 9 × 6 =
e. 8 × 6 =	k. 4 × 5 =	q. 4 × 7 =	w. 8 × 2 =	C. 5 × 2 =
f. 0 × 10 =	l. 7 × 8 =	r. 5 × 6 =	x. 7 × 7 =	D. 8 × 4 =

Copyright © The McGraw-Hill Companies, Inc.

Lesson 74

a. $6 \times 6 =$ g. $3 \times 7 =$ m. $4 \times 9 =$ s. $3 \times 2 =$ y. $3 \times 4 =$

b. $3 \times 5 =$ h. $7 \times 8 =$ n. $7 \times 6 =$ t. $6 \times 4 =$ z. $9 \times 2 =$

c. $8 \times 9 =$ i. $0 \times 9 =$ o. $2 \times 7 =$ u. $2 \times 5 =$ A. $1 \times 10 =$

d. $6 \times 8 =$ j. $7 \times 7 =$ p. $3 \times 3 =$ v. $9 \times 9 =$ B. $4 \times 6 =$

e. $5 \times 7 =$ k. $4 \times 5 =$ q. $8 \times 7 =$ w. $3 \times 6 =$ C. $1 \times 1 =$

f. $3 \times 6 =$ l. $8 \times 8 =$ r. $4 \times 8 =$ x. $4 \times 7 =$ D. $3 \times 8 =$

Part 2

Fraction	$\overline{}$	$\div$	Mixed or Whole Number	
a.		$85 \div 4$		
b.			$19\frac{1}{2}$	
c. $\frac{126}{9}$				
d.			$45\frac{3}{4}$	

Part 3

a. $5 \times 5 =$ g. $4 \times 7 =$ m. $18 \div 3 =$ s. $4 \times 8 =$ y. $30 \div 5 =$

b. $8 \times 6 =$ h. $16 \div 2 =$ n. $8 \div 4 =$ t. $28 \div 4 =$ z. $0 \div 10 =$

c. $56 \div 7 =$ i. $24 \div 4 =$ o. $6 \times 6 =$ u. $54 \div 6 =$ A. $6 \times 4 =$

d. $7 \times 7 =$ j. $42 \div 6 =$ p. $10 \div 1 =$ v. $3 \times 8 =$ B. $27 \div 3 =$

e. $0 \div 5 =$ k. $24 \div 3 =$ q. $81 \div 9 =$ w. $8 \div 8 =$ C. $32 \div 4 =$

f. $64 \div 8 =$ l. $16 \div 4 =$ r. $7 \times 3 =$ x. $8 \times 9 =$ D. $6 \times 7 =$

Lesson 75

Part 1

a. $8 \times 7 =$	g. $2 \times 4 =$	m. $9 \times 9 =$	s. $2 \times 9 =$	y. $5 \times 9 =$
b. $1 \times 4 =$	h. $6 \times 7 =$	n. $2 \times 7 =$	t. $8 \times 5 =$	z. $3 \times 2 =$
c. $9 \times 7 =$	i. $3 \times 8 =$	o. $4 \times 6 =$	u. $9 \times 6 =$	A. $7 \times 8 =$
d. $3 \times 6 =$	j. $4 \times 7 =$	p. $0 \times 8 =$	v. $4 \times 8 =$	B. $10 \times 7 =$
e. $6 \times 8 =$	k. $8 \times 8 =$	q. $8 \times 6 =$	w. $6 \times 6 =$	C. $8 \times 9 =$
f. $3 \times 3 =$	l. $3 \times 9 =$	r. $4 \times 10 =$	x. $3 \times 7 =$	D. $7 \times 6 =$

Part 2

a. $\quad 7\frac{8}{9}$
$\quad - 2\frac{3}{9}$

b. $\quad 7\frac{4}{9}$
$\quad + 2\frac{3}{9}$

c. $\quad 43\frac{6}{11}$
$\quad + 56\frac{3}{11}$

d. $\quad 98\frac{3}{5}$
$\quad - 63\frac{1}{5}$

e. $\quad 4\frac{3}{8}$
$\quad + 36\frac{2}{8}$

Part 3

Fraction	$\overline{\big)}$	$\div$	Mixed or Whole Number	
a. $\frac{84}{5}$	$\overline{\big)}$			
b.	$\overline{\big)}$		$23\frac{5}{9}$	
c.	$\overline{\big)}$	$172 \div 4$		

Lesson 75

Part 4

a. 28 ÷ 4 =	g. 56 ÷ 7 =	m. 16 ÷ 4 =	s. 30 ÷ 5 =	y. 54 ÷ 6 =
b. 49 ÷ 7 =	h. 3 × 8 =	n. 3 × 7 =	t. 24 ÷ 6 =	z. 4 × 7 =
c. 7 × 6 =	i. 72 ÷ 8 =	o. 4 × 8 =	u. 6 × 6 =	A. 64 ÷ 8 =
d. 8 ÷ 4 =	j. 7 ÷ 7 =	p. 27 ÷ 3 =	v. 36 ÷ 9 =	B. 8 × 10 =
e. 48 ÷ 6 =	k. 8 × 8 =	q. 42 ÷ 7 =	w. 24 ÷ 3 =	C. 9 × 7 =
f. 7 × 8 =	l. 32 ÷ 4 =	r. 5 × 5 =	x. 18 ÷ 6 =	D. 0 ÷ 7 =

Lesson 76

Part 1

a. 6 × 8 =	g. 6 × 9 =	m. 8 × 6 =	s. 3 × 6 =	y. 9 × 7 =
b. 2 × 6 =	h. 3 × 0 =	n. 4 × 10 =	t. 8 × 7 =	z. 2 × 8 =
c. 4 × 7 =	i. 3 × 8 =	o. 4 × 6 =	u. 4 × 4 =	A. 4 × 8 =
d. 1 × 8 =	j. 6 × 7 =	p. 3 × 9 =	v. 8 × 0 =	B. 3 × 7 =
e. 7 × 3 =	k. 1 × 1 =	q. 7 × 7 =	w. 9 × 8 =	C. 6 × 6 =
f. 8 × 8 =	l. 7 × 8 =	r. 3 × 4 =	x. 9 × 9 =	D. 4 × 5 =

Part 2

a. 20% =	b. 17% =	c. 81% =	d. 1% =

Part 3

a. $34\frac{2}{6}$
   $+\ 27\frac{3}{6}$

b. $19\frac{5}{8}$
   $-\ 6\frac{5}{8}$

c. $43\frac{8}{11}$
   $-\ 42\frac{5}{11}$

d. $94\frac{5}{9}$
   $+\ 7\frac{3}{9}$

e. $38\frac{9}{10}$
   $-\ 19\frac{3}{10}$

Lesson 76

Part 4

a. 48 ÷ 8 =	g. 3 × 7 =	m. 42 ÷ 6 =	s. 8 × 8 =	y. 18 ÷ 6 =
b. 6 × 6 =	h. 12 ÷ 4 =	n. 3 × 9 =	t. 36 ÷ 6 =	z. 64 ÷ 8 =
c. 42 ÷ 7 =	i. 56 ÷ 8 =	o. 8 × 7 =	u. 24 ÷ 3 =	A. 4 × 8 =
d. 3 × 6 =	j. 54 ÷ 6 =	p. 18 ÷ 9 =	v. 49 ÷ 7 =	B. 7 × 6 =
e. 28 ÷ 7 =	k. 4 × 1 =	q. 7 ÷ 7 =	w. 0 ÷ 9 =	C. 56 ÷ 7 =
f. 32 ÷ 8 =	l. 16 ÷ 4 =	r. 7 × 7 =	x. 6 × 8 =	D. 21 ÷ 3 =

Lesson 77

Part 1

a. 9 × 10 =	g. 7 × 6 =	m. 6 × 7 =	s. 8 × 7 =	y. 9 × 6 =
b. 7 × 4 =	h. 5 × 5 =	n. 1 × 3 =	t. 4 × 4 =	z. 9 × 9 =
c. 3 × 5 =	i. 6 × 6 =	o. 9 × 3 =	u. 3 × 6 =	A. 7 × 1 =
d. 6 × 4 =	j. 6 × 3 =	p. 4 × 5 =	v. 7 × 9 =	B. 3 × 8 =
e. 8 × 8 =	k. 0 × 10 =	q. 6 × 8 =	w. 2 × 9 =	C. 5 × 2 =
f. 7 × 8 =	l. 4 × 8 =	r. 3 × 4 =	x. 7 × 7 =	D. 10 × 10 =

Part 2

a. 20%	c. 5%	e. 1%
b. 320%	d. 400%	f. 708%

Part 3

a. 56 ÷ 8 =	g. 49 ÷ 7 =	m. 8 × 3 =	s. 42 ÷ 6 =	y. 7 × 7 =
b. 48 ÷ 6 =	h. 0 ÷ 10 =	n. 10 ÷ 10 =	t. 4 × 9 =	z. 8 ÷ 2 =
c. 32 ÷ 8 =	i. 6 × 6 =	o. 21 ÷ 7 =	u. 7 × 8 =	A. 36 ÷ 6 =
d. 21 ÷ 3 =	j. 9 × 1 =	p. 54 ÷ 9 =	v. 64 ÷ 8 =	B. 4 × 8 =
e. 6 × 7 =	k. 42 ÷ 7 =	q. 28 ÷ 4 =	w. 8 × 8 =	C. 16 ÷ 2 =
f. 18 ÷ 3 =	l. 16 ÷ 4 =	r. 1 × 3 =	x. 10 ÷ 5 =	D. 12 ÷ 4 =

Lesson

a. 7 × 7 =   g. 3 × 7 =   m. 4 × 3 =   s. 6 × 6 =   y. 2 × 7 =
b. 3 × 5 =   h. 8 × 8 =   n. 4 × 7 =   t. 5 × 7 =   z. 3 × 9 =
c. 4 × 9 =   i. 9 × 6 =   o. 8 × 3 =   u. 4 × 6 =   A. 5 × 6 =
d. 8 × 7 =   j. 1 × 5 =   p. 8 × 9 =   v. 7 × 9 =   B. 3 × 10 =
e. 3 × 6 =   k. 4 × 8 =   q. 0 × 8 =   w. 4 × 4 =   C. 9 × 5 =
f. 6 × 8 =   l. 7 × 6 =   r. 3 × 8 =   x. 10 × 8 =   D. 7 × 8 =

Part 2

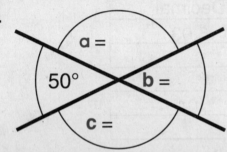

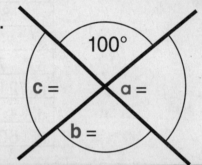

Part 3

a. 24 ÷ 3 =   g. 18 ÷ 3 =   m. 48 ÷ 6 =   s. 16 ÷ 4 =   y. 4 × 8 =
b. 72 ÷ 8 =   h. 54 ÷ 6 =   n. 63 ÷ 7 =   t. 6 × 6 =   z. 36 ÷ 4 =
c. 8 × 6 =   i. 1 × 8 =   o. 0 ÷ 2 =   u. 12 ÷ 3 =   A. 6 × 7 =
d. 56 ÷ 7 =   j. 42 ÷ 6 =   p. 7 × 7 =   v. 21 ÷ 3 =   B. 32 ÷ 4 =
e. 9 × 9 =   k. 64 ÷ 8 =   q. 36 ÷ 6 =   w. 7 × 8 =   C. 8 ÷ 8 =
f. 24 ÷ 4 =   l. 0 × 8 =   r. 28 ÷ 7 =   x. 49 ÷ 7 =   D. 8 × 8 =

Lesson

Part 1

a. 8 × 4 =   g. 3 × 8 =   m. 4 × 6 =   s. 6 × 6 =   y. 7 × 1 =
b. 1 × 9 =   h. 3 × 4 =   n. 7 × 9 =   t. 2 × 10 =   z. 8 × 8 =
c. 2 × 3 =   i. 8 × 6 =   o. 0 × 10 =   u. 3 × 7 =   A. 6 × 4 =
d. 3 × 7 =   j. 1 × 0 =   p. 4 × 9 =   v. 6 × 9 =   B. 5 × 8 =
e. 7 × 6 =   k. 4 × 8 =   q. 6 × 3 =   w. 8 × 7 =   C. 4 × 7 =
f. 9 × 9 =   l. 7 × 7 =   r. 3 × 3 =   x. 4 × 4 =   D. 10 × 1 =

Lesson 79

Part 2

1.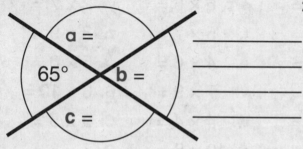

   _____
   _____
   _____
   _____

2.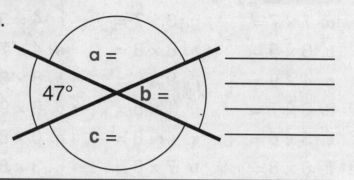

   _____
   _____
   _____
   _____

Part 3

	%	Decimal
a.		6.03
b.	7%	
c.	216%	
d.		.04
e.	500%	

Part 4

a. $24 \div 8 =$ g. $3 \times 7 =$ m. $7 \times 8 =$ s. $7 \times 7 =$ y. $18 \div 9 =$

b. $48 \div 6 =$ h. $16 \div 4 =$ n. $49 \div 7 =$ t. $4 \times 8 =$ z. $32 \div 8 =$

c. $4 \times 9 =$ i. $8 \div 8 =$ o. $2 \times 1 =$ u. $45 \div 5 =$ A. $56 \div 8 =$

d. $42 \div 6 =$ j. $0 \div 3 =$ p. $12 \div 4 =$ v. $18 \div 3 =$ B. $54 \div 9 =$

e. $7 \times 4 =$ k. $24 \div 6 =$ q. $15 \div 3 =$ w. $6 \times 8 =$ C. $7 \times 6 =$

f. $56 \div 7 =$ l. $9 \times 0 =$ r. $36 \div 6 =$ x. $64 \div 8 =$ D. $10 \times 10 =$

Lesson 80

Part 1

a. 2 ⌐6→ ___   d. 6 ⌐8→ ___   g. ___ ⌐6→ 24   j. ___ ⌐8→ 64

b. 7 ⌐→ 28   e. 9 ⌐8→ ___   h. 7 ⌐→ 56   k. 3 ⌐7→ ___

c. ___ ⌐9→ 27   f. 7 ⌐→ 49   i. 4 ⌐8→ ___   l. 6 ⌐→ 42

Connecting Math Concepts

Lesson 80

Part 2

a. $63 \div 9 =$	g. $24 \div 3 =$	m. $28 \div 7 =$	s. $4 \div 4 =$	y. $36 \div 6 =$
b. $16 \div 8 =$	h. $9 \div 1 =$	n. $72 \div 9 =$	t. $64 \div 8 =$	z. $56 \div 8 =$
c. $42 \div 6 =$	i. $56 \div 7 =$	o. $21 \div 3 =$	u. $36 \div 9 =$	A. $24 \div 6 =$
d. $32 \div 4 =$	j. $54 \div 9 =$	p. $56 \div 7 =$	v. $28 \div 4 =$	B. $12 \div 2 =$
e. $0 \div 6 =$	k. $64 \div 8 =$	q. $24 \div 4 =$	w. $80 \div 10 =$	C. $18 \div 9 =$
f. $49 \div 7 =$	l. $36 \div 4 =$	r. $48 \div 8 =$	x. $81 \div 9 =$	D. $27 \div 3 =$

Part 3

a. $3\frac{4}{9} = 2 + \underline{\quad}$

b. $7\frac{3}{8} = 6 + \underline{\quad}$

c. $9\frac{1}{2} = 8 + \underline{\quad}$

d. $3\frac{2}{5} = 2 + \underline{\quad}$

Part 4

	%	Decimal
a.	800%	
b.		.05
c.	127%	
d.	4%	
e.		3.60

Part 5

a. $7 \times 9 =$	g. $4 \times 9 =$	m. $3 \times 9 =$	s. $9 \times 7 =$	y. $5 \times 7 =$
b. $8 \times 6 =$	h. $7 \times 8 =$	n. $8 \times 7 =$	t. $6 \times 4 =$	z. $6 \times 8 =$
c. $7 \times 3 =$	i. $7 \times 6 =$	o. $6 \times 10 =$	u. $8 \times 4 =$	A. $7 \times 7 =$
d. $2 \times 6 =$	j. $8 \times 5 =$	p. $4 \times 3 =$	v. $5 \times 7 =$	B. $2 \times 5 =$
e. $3 \times 4 =$	k. $4 \times 7 =$	q. $8 \times 3 =$	w. $3 \times 0 =$	C. $6 \times 3 =$
f. $8 \times 8 =$	l. $3 \times 8 =$	r. $9 \times 8 =$	x. $6 \times 1 =$	D. $9 \times 9 =$

Lesson

a. $8 \times 8 =$ g. $3 \times 7 =$ m. $0 \times 9 =$ s. $5 \times 8 =$ y. $8 \times 4 =$

b. $4 \times 5 =$ h. $5 \times 7 =$ n. $9 \times 6 =$ t. $6 \times 7 =$ z. $4 \times 10 =$

c. $4 \times 8 =$ i. $6 \times 6 =$ o. $7 \times 9 =$ u. $3 \times 5 =$ A. $2 \times 2 =$

d. $3 \times 6 =$ j. $4 \times 7 =$ p. $3 \times 4 =$ v. $8 \times 0 =$ B. $4 \times 9 =$

e. $5 \times 6 =$ k. $6 \times 8 =$ q. $8 \times 6 =$ w. $7 \times 7 =$ C. $1 \times 9 =$

f. $4 \times 6 =$ l. $8 \times 3 =$ r. $4 \times 4 =$ x. $9 \times 8 =$ D. $7 \times 8 =$

Part 2

a.
$$
\begin{array}{r}
3\,5\,3 \\
-\ 8\,9 \\
\hline
\end{array}
$$

b.
$$
\begin{array}{r}
1\,7 \\
8\,4 \\
1\,0\,2 \\
+\ 7\,9 \\
\hline
\end{array}
$$

c.
$$
\begin{array}{r}
7\,2\,8 \\
-6\,8\,1 \\
\hline
\end{array}
$$

Part 3

a. $8 \times 1 =$ g. $32 \div 8 =$ m. $18 \div 6 =$ s. $4 \times 7 =$ y. $24 \div 3 =$

b. $0 \div 2 =$ h. $12 \div 4 =$ n. $1 \div 1 =$ t. $42 \div 7 =$ z. $7 \times 7 =$

c. $48 \div 6 =$ i. $21 \div 7 =$ o. $5 \times 0 =$ u. $8 \times 8 =$ A. $28 \div 7 =$

d. $6 \times 7 =$ j. $4 \times 6 =$ p. $8 \times 6 =$ v. $18 \div 3 =$ B. $3 \times 6 =$

e. $49 \div 7 =$ k. $90 \div 9 =$ q. $63 \div 9 =$ w. $45 \div 9 =$ C. $72 \div 9 =$

f. $6 \times 6 =$ l. $56 \div 8 =$ r. $7 \times 8 =$ x. $64 \div 8 =$ D. $36 \div 6 =$

Lesson

Part 1

a. $7 \times 7 =$	g. $7 \times 4 =$	m. $1 \times 5 =$	s. $10 \times 2 =$	y. $9 \times 5 =$
b. $8 \times 8 =$	h. $3 \times 6 =$	n. $9 \times 4 =$	t. $3 \times 9 =$	z. $10 \times 7 =$
c. $4 \times 6 =$	i. $1 \times 0 =$	o. $7 \times 8 =$	u. $4 \times 8 =$	A. $5 \times 6 =$
d. $6 \times 9 =$	j. $3 \times 8 =$	p. $8 \times 9 =$	v. $9 \times 9 =$	B. $6 \times 8 =$
e. $3 \times 5 =$	k. $7 \times 6 =$	q. $7 \times 3 =$	w. $3 \times 4 =$	C. $4 \times 7 =$
f. $6 \times 6 =$	l. $8 \times 6 =$	r. $8 \times 0 =$	x. $6 \times 2 =$	D. $8 \times 7 =$

Independent Work

Part 2

a.
$$\begin{array}{r} 37.08 \\ -5.3 \\ \hline \end{array}$$

c.
$$\begin{array}{r} 6 \\ +18.5 \\ \hline \end{array}$$

e.
$$\begin{array}{r} 7 \\ 7.84 \\ +83.2 \\ \hline \end{array}$$

b.
$$\begin{array}{r} 52.71 \\ 28 \\ +3.1 \\ \hline \end{array}$$

d.
$$\begin{array}{r} 47.4 \\ -4.74 \\ \hline \end{array}$$

Part 3

a. $56 \div 7 =$	g. $9 \times 1 =$	m. $16 \div 4 =$	s. $18 \div 6 =$	y. $63 \div 9 =$
b. $12 \div 3 =$	h. $0 \div 9 =$	n. $32 \div 4 =$	t. $24 \div 8 =$	z. $1 \times 7 =$
c. $6 \times 7 =$	i. $48 \div 8 =$	o. $7 \times 7 =$	u. $54 \div 6 =$	A. $4 \times 7 =$
d. $36 \div 4 =$	j. $6 \times 8 =$	p. $3 \times 6 =$	v. $42 \div 7 =$	B. $21 \div 7 =$
e. $49 \div 7 =$	k. $50 \div 10 =$	q. $0 \times 9 =$	w. $6 \times 10 =$	C. $36 \div 6 =$
f. $3 \times 8 =$	l. $24 \div 6 =$	r. $18 \div 9 =$	x. $64 \div 8 =$	D. $9 \times 4 =$

Lesson

a. $6 \times 0 =$ g. $3 \times 8 =$ m. $3 \times 4 =$ s. $1 \times 8 =$ y. $4 \times 8 =$

b. $7 \times 6 =$ h. $7 \times 4 =$ n. $8 \times 8 =$ t. $9 \times 8 =$ z. $9 \times 9 =$

c. $3 \times 6 =$ i. $5 \times 5 =$ o. $4 \times 9 =$ u. $4 \times 4 =$ A. $3 \times 3 =$

d. $9 \times 7 =$ j. $7 \times 8 =$ p. $6 \times 6 =$ v. $0 \times 6 =$ B. $8 \times 6 =$

e. $6 \times 8 =$ k. $9 \times 1 =$ q. $3 \times 7 =$ w. $8 \times 7 =$ C. $5 \times 7 =$

f. $9 \times 6 =$ l. $8 \times 4 =$ r. $7 \times 7 =$ x. $6 \times 4 =$ D. $9 \times 3 =$

Part 2

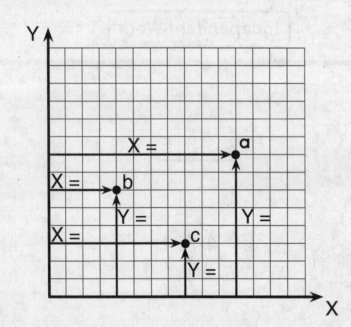

Part 3

a. $8 \times 3 =$ g. $24 \div 3 =$ m. $56 \div 8 =$ s. $24 \div 6 =$ y. $4 \times 9 =$

b. $18 \div 6 =$ h. $16 \div 4 =$ n. $42 \div 6 =$ t. $4 \times 7 =$ z. $64 \div 8 =$

c. $7 \times 7 =$ i. $6 \times 6 =$ o. $12 \div 3 =$ u. $16 \div 2 =$ A. $36 \div 6 =$

d. $1 \times 5 =$ j. $36 \div 9 =$ p. $32 \div 4 =$ v. $3 \times 6 =$ B. $8 \div 8 =$

e. $48 \div 8 =$ k. $42 \div 6 =$ q. $6 \times 7 =$ w. $72 \div 8 =$ C. $7 \times 3 =$

f. $8 \times 0 =$ l. $8 \times 8 =$ r. $81 \div 9 =$ x. $49 \div 7 =$ D. $40 \div 8 =$

Lesson

Part 1

a. 9 × 9 =	g. 1 × 10 =	m. 3 × 8 =	s. 9 × 7 =	y. 8 × 3 =
b. 7 × 8 =	h. 4 × 6 =	n. 4 × 3 =	t. 6 × 4 =	z. 4 × 10 =
c. 4 × 7 =	i. 3 × 2 =	o. 6 × 6 =	u. 5 × 3 =	A. 1 × 1 =
d. 2 × 6 =	j. 3 × 7 =	p. 4 × 9 =	v. 8 × 8 =	B. 9 × 6 =
e. 8 × 6 =	k. 7 × 6 =	q. 6 × 5 =	w. 4 × 4 =	C. 7 × 7 =
f. 3 × 6 =	l. 8 × 4 =	r. 8 × 7 =	x. 9 × 1 =	D. 3 × 9 =

Part 2

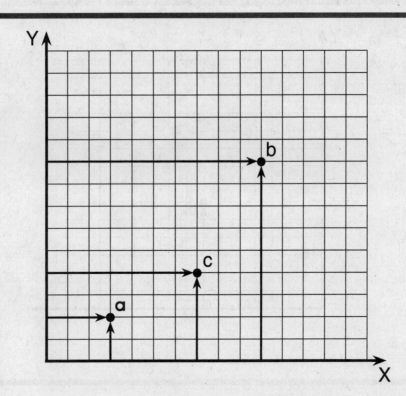

Part 3

a. 36 ÷ 6 =	g. 24 ÷ 6 =	m. 7 × 8 =	s. 72 ÷ 8 =	y. 7 × 3 =
b. 56 ÷ 7 =	h. 48 ÷ 6 =	n. 16 ÷ 4 =	t. 1 × 9 =	z. 42 ÷ 6 =
c. 4 × 7 =	i. 70 ÷ 7 =	o. 7 × 0 =	u. 21 ÷ 7 =	A. 63 ÷ 7 =
d. 64 ÷ 8 =	j. 0 ÷ 3 =	p. 18 ÷ 3 =	v. 54 ÷ 6 =	B. 7 × 6 =
e. 24 ÷ 8 =	k. 6 × 7 =	q. 32 ÷ 4 =	w. 18 ÷ 6 =	C. 49 ÷ 7 =
f. 9 × 3 =	l. 32 ÷ 8 =	r. 7 × 7 =	x. 5 ÷ 5 =	D. 6 × 9 =

Lesson

a. 4 × 9 =   g. 7 × 7 =   m. 4 × 6 =   s. 7 × 6 =   y. 3 × 2 =

b. 6 × 8 =   h. 8 × 5 =   n. 3 × 1 =   t. 4 × 5 =   z. 4 × 7 =

c. 3 × 7 =   i. 4 × 4 =   o. 4 × 8 =   u. 7 × 8 =   A. 1 × 1 =

d. 8 × 8 =   j. 0 × 10 =   p. 6 × 3 =   v. 6 × 9 =   B. 7 × 3 =

e. 2 × 7 =   k. 8 × 7 =   q. 9 × 7 =   w. 1 × 8 =   C. 6 × 6 =

f. 4 × 3 =   l. 6 × 7 =   r. 3 × 8 =   x. 5 × 9 =   D. 9 × 8 =

Part 2

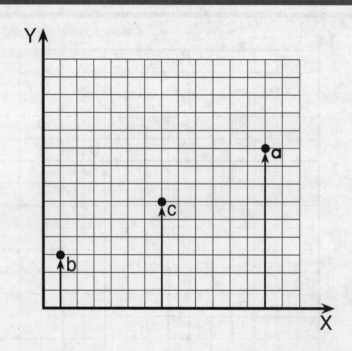

Part 3

a. 1 × 2 =   g. 32 ÷ 8 =   m. 18 ÷ 3 =   s. 9 × 4 =   y. 49 ÷ 7 =

b. 8 × 6 =   h. 48 ÷ 6 =   n. 8 × 4 =   t. 15 ÷ 3 =   z. 0 × 9 =

c. 16 ÷ 4 =   i. 28 ÷ 7 =   o. 56 ÷ 7 =   u. 21 ÷ 7 =   A. 36 ÷ 6 =

d. 42 ÷ 6 =   j. 8 × 3 =   p. 6 × 6 =   v. 28 ÷ 4 =   B. 7 × 8 =

e. 7 × 4 =   k. 36 ÷ 6 =   q. 12 ÷ 3 =   w. 7 × 1 =   C. 24 ÷ 4 =

f. 27 ÷ 9 =   l. 8 × 7 =   r. 0 ÷ 9 =   x. 64 ÷ 8 =   D. 8 × 8 =

Connecting Math Concepts

Lesson 86

Part 1

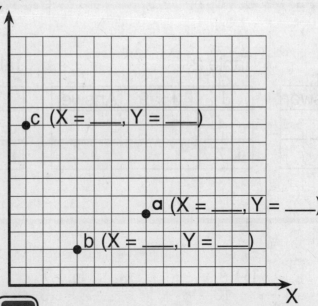

c ● (X = ____ , Y = ____)

a ● (X = ____ , Y = ____)

b ● (X = ____ , Y = ____)

Lesson 87

Part 1

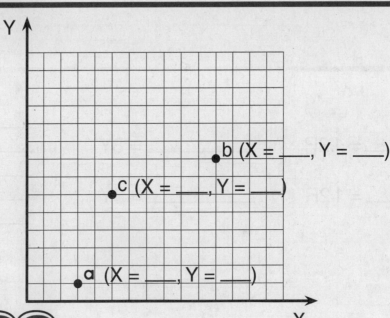

b ● (X = ____ , Y = ____)

c ● (X = ____ , Y = ____)

a ● (X = ____ , Y = ____)

Lesson 88

Part 1

	X	X + 5	Answer
a.	7		
b.	4		
c.	25		

Lesson

Table 1

V	V − 6	Answer
a. 10		
b. 15		

Table 2

J	J × 9	Answer
c. 4		
d. 7		

Table 3

Y	Y + 13	Answer
e. 0		
f. 4		

Part 2

a. $\begin{array}{r} 8 \\ -\ 6\frac{3}{100} \\ \hline \end{array}$

b. $\begin{array}{r} 19\frac{4}{10} \\ +\ 4\frac{6}{10} \\ \hline \end{array}$

c. $\begin{array}{r} 53\frac{7}{9} \\ +\ 62\frac{1}{9} \\ \hline \end{array}$

d. $\begin{array}{r} 46\frac{9}{20} \\ -\ 9\frac{5}{20} \\ \hline \end{array}$

Part 3

a. _____ = 1.2R

_____ = 1.2R

b. _____ = 8Y

_____ = 8Y

c. $\dfrac{\rule{3cm}{0.4pt}}{\rule{3cm}{0.4pt}} = \frac{3}{2}G$

$\dfrac{\rule{3cm}{0.4pt}}{\rule{3cm}{0.4pt}} = \frac{3}{2}G$

Lesson 90

Table 1

R	39 − R	Answer
a. 10		
b. 3		

Table 2

G	3G	Answer
c. 5		
d. 9		

Table 3

X	X ÷ 2	Answer
e. 20		
f. 8		

Part 2

a. $60\frac{7}{10}$

$+\ 28\frac{6}{10}$

b. $52\frac{5}{8}$

$-\ 46$

c. 35

$-\ 19\frac{2}{5}$

d. $74\frac{13}{100}$

$+\ 38\frac{5}{100}$

Lesson 91

Part 1

Table 1

M	5M	Answer
a. 1		
b. 8		

Table 2

Q	Q − 23	Answer
c. 33		
d. 53		

Table 3

R	70 ÷ R	Answer
e. 10		
f. 7		

Lesson

Part 1

M	Function 3M	P
a. 1		
b. 0		
c. 10		
d. 30		

Part 2

_____ (X = 5, Y = 4)

_____ (X = 7, Y = ___)

_____ (X = ___, Y = 7)

_____ (X = 1, Y = ___)

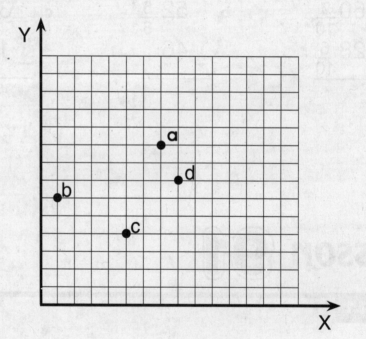

Connecting Math Concepts

Lesson 93

Part 1

Table 1

	Function	
F	4F	Y
a. 6		
b. 2		
c. 5		
d. 1		

Table 2

	Function	
N	N − 3	Q
a. 3		
b. 10		
c. 13		
d. 47		

Lesson 94

Part 1

	Function	
X		Y
___. 2		
___.		3
___. 4		
___. 7		
___.		9

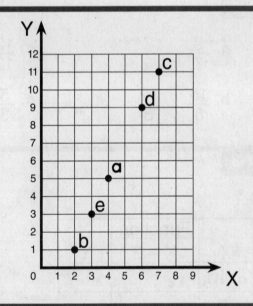

Part 2

	Function	
X	2X	Y
a. 6		
b. 2		
c. 5		
d. 1		
e.		

Lesson 95

Part 1

X	Function	Y
___. 4		
___.		6
___.		9
___. 10		

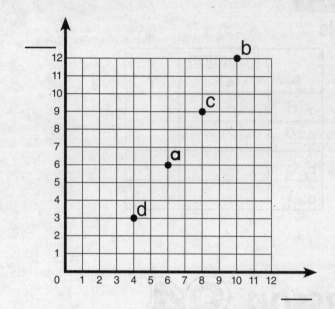

Part 2

a. $\dfrac{3}{7}$ $\dfrac{7}{3}$ c. $\dfrac{12}{12}$ $\dfrac{11}{8}$ e. $\dfrac{14}{9}$ $\dfrac{15}{16}$

b. $\dfrac{15}{8}$ $\dfrac{12}{8}$ d. $\dfrac{5}{3}$ $\dfrac{4}{3}$ f. $\dfrac{6}{10}$ $\dfrac{9}{10}$

Part 3

Table 1

X	Function X − 3	Y
a. 10		
b. 7		
c. 4		

Table 2

X	Function 9 − X	Y
d. 9		
e. 4		
f. 0		

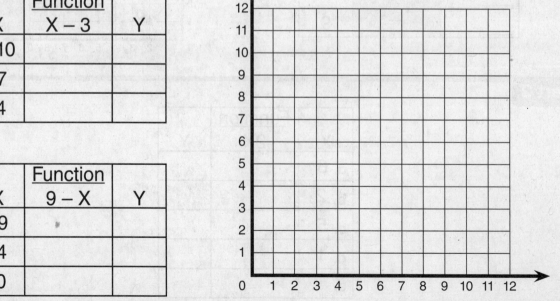

Connecting Math Concepts

Lesson 96

Part 1

X	Function	Y
___.4		
___.		5
___.		6
___.6		

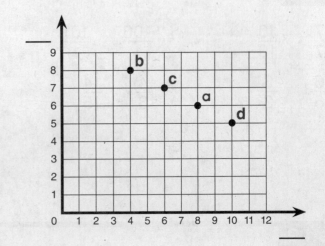

Part 2

a. $\dfrac{4}{4}$ $\dfrac{1}{1}$ c. $\dfrac{29}{5}$ $\dfrac{31}{5}$ e. $\dfrac{8}{8}$ $\dfrac{7}{5}$

b. $\dfrac{15}{14}$ $\dfrac{9}{9}$ d. $\dfrac{12}{11}$ $\dfrac{13}{14}$ f. $\dfrac{21}{47}$ $\dfrac{19}{47}$

Part 3

Table 1

X	Function 3X	Y
a. 0		
b. 4		
c. 2		

Table 2

X	Function X + 3	Y
d. 10		
e. 7		
f. 0		

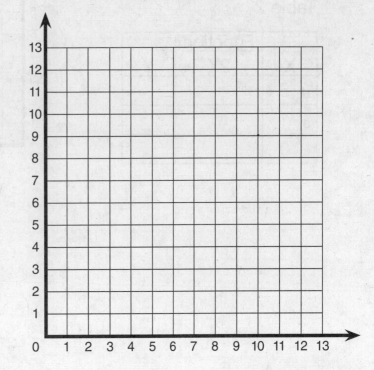

Lesson

a. $\dfrac{7}{7}$ $\dfrac{10}{10}$ c. $\dfrac{100}{1}$ $\dfrac{101}{1}$ e. $\dfrac{3}{10}$ $\dfrac{1}{10}$ g. $\dfrac{25}{25}$ $\dfrac{9}{9}$

b. $\dfrac{6}{5}$ $\dfrac{3}{3}$ d. $\dfrac{4}{3}$ $\dfrac{14}{15}$ f. $\dfrac{6}{6}$ $\dfrac{19}{20}$ h. $\dfrac{7}{24}$ $\dfrac{8}{24}$

Part 2

Table 1

	Function	
X	11 − X	Y
a. 0		
b. 6		
c. 11		

Table 2

	Function	
X	2X	Y
d. 6		
e. 1		
f. 0		

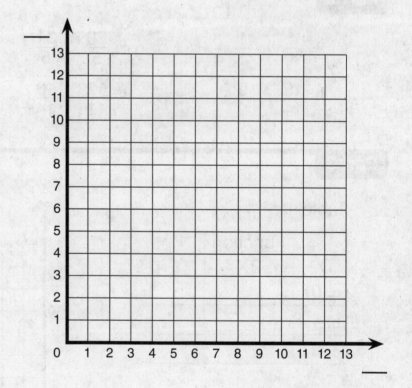

Lesson

X	Function $\frac{5}{2}X$	Y
a. 2		
b. 3		
c. 4		
d. 5		

Part 2

a. $\frac{3}{6}$ $\frac{3}{5}$ e. $\frac{4}{4}$ $\frac{8}{9}$

b. $\frac{3}{5}$ $\frac{2}{5}$ f. $\frac{8}{11}$ $\frac{8}{15}$

c. $\frac{14}{10}$ $\frac{14}{9}$ g. $\frac{12}{12}$ $\frac{5}{5}$

d. $\frac{14}{15}$ $\frac{13}{12}$ h. $\frac{9}{2}$ $\frac{9}{3}$

Part 3

a. $\quad 4\frac{2}{9}$
   $\quad \times \ 8$

b. $\quad 24\frac{2}{3}$
   $\quad \times \ 7$

Lesson

Table 1

J	Function $\frac{3}{4}$J	Q
a. 0		
b. 4		
c. 5		
d. 8		

Table 2

R	Function $\frac{5}{3}$R	M
a. 1		
b. 3		
c. 5		
d. 9		

Part 2

a. $39\frac{4}{7}$
$\times\ \ 5$

b. $16\frac{3}{5}$
$\times\ \ 9$

Part 3

X	Function 3 + X	Y
a. 7		10
b. 0		3
c. 2		
d.		4
e. 5		

Lesson 100

Part 1

X	Function $\frac{3}{2}$ X	Y
a. 2		
b. 4		
c. 6		
d. 8		
e. 10		

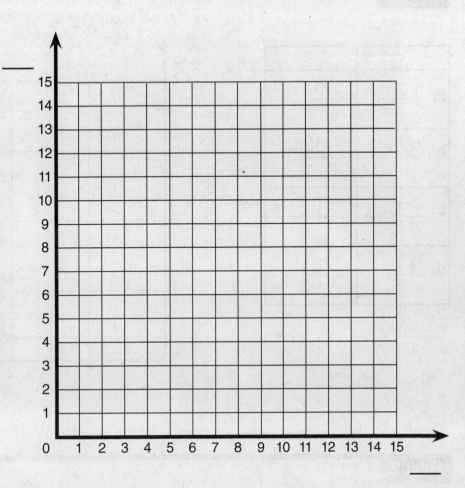

Part 2

a. $\frac{9}{9}$ $\frac{4}{3}$

b. $\frac{4}{5}$ $\frac{4}{3}$

c. $\frac{5}{2}$ $\frac{5}{4}$

d. $\frac{17}{17}$ $\frac{2}{2}$

e. $\frac{9}{5}$ $\frac{8}{5}$

f. $\frac{11}{20}$ $\frac{11}{12}$

g. $\frac{14}{14}$ $\frac{17}{18}$

h. $\frac{26}{100}$ $\frac{31}{100}$

i. $\frac{4}{1}$ $\frac{4}{3}$

Part 1

	X	Y
a.	$1\frac{2}{3}$	2
b.	3	$2\frac{2}{3}$
c.	$\frac{1}{3}$	$1\frac{2}{3}$
d.	$3\frac{2}{3}$	$2\frac{1}{3}$

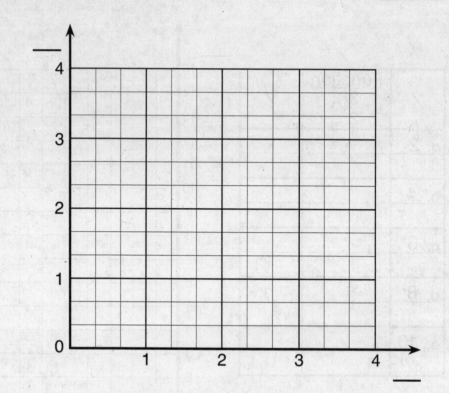

Part 2

a. 80   d. 74   f. 91

b. 35   e. 110   g. 145

c. 93

2	
5	

Lesson 102

Part 1

X	Y
a. 3	$1\frac{3}{4}$
b. $2\frac{2}{4}$	4
c. $1\frac{1}{4}$	$\frac{3}{4}$
d. $3\frac{2}{4}$	$2\frac{1}{4}$
e. $\frac{1}{4}$	$3\frac{2}{4}$

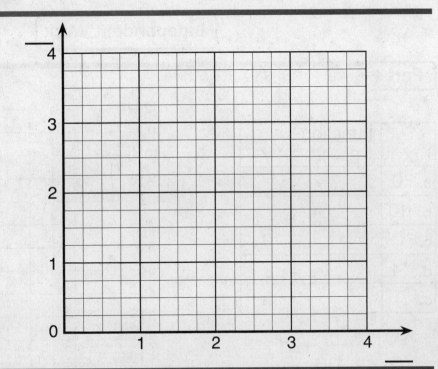

Part 2

a. 120 d. 302 g. 451

b. 407 e. 100 h. 145

c. 335 f. 154

2	
5	

Independent Work

Part 3

X	Function $\frac{3}{2}$X	Y
a. 0		
b. 6		
c. 4		
d.		3
e. 8		

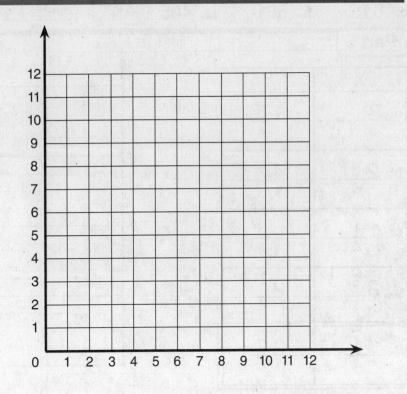

Lesson 103

Part 1

X	Function 2 + X	Y
a. 0		
b. 10		
c.		9
d. 4		
e.		3

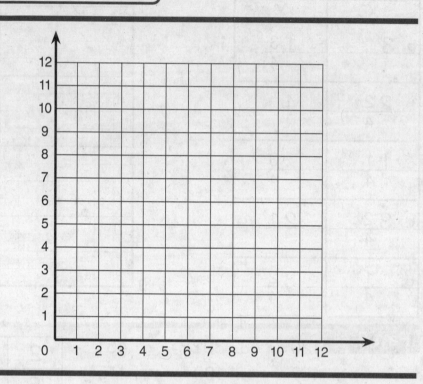

Part 2

a. 390	d. 158	g. 450
b. 71	e. 185	h. 504
c. 645	f. 581	i. 405

2	
5	

Part 3

X	Y
a. 3	$1\frac{3}{4}$
b. $2\frac{2}{4}$	4
c. $1\frac{1}{4}$	$\frac{3}{4}$
d. $3\frac{2}{4}$	$2\frac{1}{4}$
e. $\frac{1}{4}$	$3\frac{2}{4}$

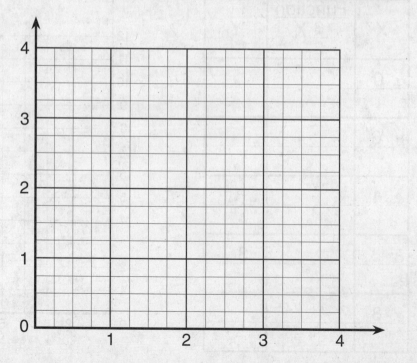

Connecting Math Concepts

Lesson 104

Part 1

a. $\frac{5}{24}$ $\frac{5}{19}$ _____

b. $\frac{27}{8}$ $\frac{8}{2}$ _____

c. $\frac{15}{12}$ $\frac{16}{17}$ _____

d. $\frac{37}{10}$ $\frac{29}{10}$ _____

e. $\frac{49}{7}$ $\frac{63}{9}$ _____

f. $\frac{94}{94}$ $\frac{6}{8}$ _____

g. $\frac{21}{3}$ $\frac{21}{2}$ _____

h. $\frac{61}{3}$ $\frac{39}{2}$ _____

Lesson 105

Part 1

a. $\frac{35}{7}$ $\frac{14}{3}$ _____

b. $\frac{32}{5}$ $\frac{32}{4}$ _____

c. $\frac{64}{65}$ $\frac{2}{2}$ _____

d. $\frac{9}{3}$ $\frac{24}{8}$ _____

e. $\frac{16}{21}$ $\frac{14}{21}$ _____

f. $\frac{26}{21}$ $\frac{34}{35}$ _____

Lesson 106

Part 1

a. $\frac{16}{15}$ $\frac{27}{27}$ _____

b. $\frac{30}{6}$ $\frac{35}{5}$ _____

c. $\frac{12}{7}$ $\frac{12}{9}$ _____

d. $\frac{16}{2}$ $\frac{40}{5}$ _____

e. $\frac{23}{19}$ $\frac{21}{19}$ _____

f. $\frac{53}{54}$ $\frac{4}{3}$ _____

g. $\frac{9}{2}$ $\frac{35}{9}$ _____

h. $\frac{37}{37}$ $\frac{5}{5}$ _____

Part 2

a. 84

b. 316

c. 425

d. 120

e. 79

f. 507

g. 705

h. 750

2	
3	
5	

Lesson 107

Part 1

a. $\dfrac{18}{3}$ $\dfrac{38}{7}$ _____

b. $\dfrac{27}{9}$ $\dfrac{27}{7}$ _____

c. $\dfrac{63}{61}$ $\dfrac{4}{5}$ _____

d. $\dfrac{2}{17}$ $\dfrac{5}{17}$ _____

e. $\dfrac{24}{3}$ $\dfrac{40}{5}$ _____

f. $\dfrac{31}{31}$ $\dfrac{2}{3}$ _____

Part 2

a. 96

b. 68

c. 540

d. 79

e. 325

f. 264

g. 251

h. 231

2	
3	
5	

Lesson 108

Part 1

a. $3\overline{)70}$

b. $3\overline{)376}$

c. $3\overline{)87}$

d. $3\overline{)270}$

e. $3\overline{)97}$

f. $3\overline{)65}$

2	
3	
5	

Lesson 109

Part 1

a. $3\overline{)240}$

c. $3\overline{)83}$

e. $2\overline{)108}$

b. $2\overline{)65}$

d. $5\overline{)105}$

f. $5\overline{)117}$

2	
3	
5	

Lesson 110

Part 1

a. $5\overline{)89}$

c. $2\overline{)78}$

e. $3\overline{)173}$

b. $3\overline{)675}$

d. $5\overline{)260}$

f. $2\overline{)183}$

2	
3	
5	

Part 2

a. 8.09   8.23

b. 12.1   11.9

c. 47.36   47.29

Connecting Math Concepts

Lesson 111

Part 1

a. 14.17   14.1   c. 32.5   32.48

b. 105   105.2   d. 9.3   9.31

Lesson 112

Part 1

a. 34.6   34.16   c. 29.10   29.1

b. 250.8   258   d. 2457.14   2458

Lesson 113

Part 1

a. 7640   7600.40   c. 87.5   87.50

b. 1206   1206.8   d. 104.3   104.03

Part 2

a. 48   c. 64   e. 35   g. 77   i. 91

b. 21   d. 84   f. 93   h. 81

7	
2	
3	
5	

Lesson 114

Part 1

Table 1

X	Function 5 + X	Y
a.		10
b.		15
c.		20

Table 2

X	Function X − 10	Y
d.		10
e.		15
f.		20

Part 2

a. $7\overline{)98}$ b. $7\overline{)89}$ c. $7\overline{)32}$ d. $7\overline{)84}$

Lesson 115

Part 1

Table 1

X	Function X + 35	Y
a.		35
b.		40
c.		45

Table 2

X	Function X − 9	Y
d.		1
e.		10
f.		22

Lesson 116

Part 1

Table 1

X	Function 3X	Y
a.		27
b.		6
c.		0

Table 2

X	Function X ÷ 5	Y
d.		5
e.		10
f.		0

Table 3

X	Function 27 − X	Y
g.		20
h.		10
i.		0

Table 4

X	Function 24 ÷ X	Y
j.		1
k.		4
l.		8

Part 2

a. hammers
   tools $\frac{3}{8}$ —— of the _____ were _____.

b. roses
   flowers $\frac{4}{9}$ —— of the _____ were _____.

c. apples
   fruit $\frac{1}{2}$ —— of the _____ were _____.

Copyright © The McGraw-Hill Companies, Inc.

Lesson 117

Part 1

a. red trucks $\frac{2}{7}$ — of the _____ were _____.

b. cars vehicles $\frac{3}{4}$ — of the _____ were _____.

c. hungry animals $\frac{5}{8}$ — of the _____ were _____.

d. frozen bags $\frac{2}{7}$ — of the _____ were _____.

Part 2

Table 1

X	Function X ÷ 2	Y
a.		20
b.		8
c.		0

Table 2

X	Function 5X	Y
d.		5
e.		35
f.		0

Table 3

X	Function X − 20	Y
g.		20
h.		10
i.		0

Table 4

X	Function 30 ÷ X	Y
j.		1
k.		3
l.		5

Connecting Math Concepts

Lesson 118

Table 1

X	Function 9X	Y
a.		27
b.		9
c.		0

Table 2

X	Function X ÷ 2	Y
d.		10
e.		50
f.		0

Table 3

X	Function X − 100	Y
g.		100
h.		25
i.		10

Table 4

X	Function 100 ÷ X	Y
j.		100
k.		25
l.		10

Part 2

a. 63   c. 56   e. 84   g. 15

b. 80   d. 41   f. 65   h. 91

2	
3	
5	
7	

Lesson 119

a. 42   c. 50   e. 23   g. 84

b. 19   d. 63   f. 85   h. 35

2	
3	
5	
7	

Part 2

a. $\dfrac{5}{7} \times \dfrac{12}{12} = \dfrac{60}{84}$
$\dfrac{5}{7} \quad \dfrac{60}{84}$

c. $\dfrac{2}{3} \times \dfrac{15}{16} = \dfrac{30}{48}$
$\dfrac{2}{3} \quad \dfrac{30}{48}$

e. $\dfrac{6}{5} \times \dfrac{15}{15} = \dfrac{90}{75}$
$\dfrac{6}{5} \quad \dfrac{90}{75}$

b. $\dfrac{3}{2} \times \dfrac{28}{27} = \dfrac{84}{54}$
$\dfrac{3}{2} \quad \dfrac{84}{54}$

d. $\dfrac{4}{9} \times \dfrac{21}{21} = \dfrac{84}{189}$
$\dfrac{4}{9} \quad \dfrac{84}{189}$

f. $\dfrac{1}{8} \times \dfrac{16}{15} = \dfrac{16}{120}$
$\dfrac{1}{8} \quad \dfrac{16}{120}$

Part 3

Table 1

X	Function 3X	Y
a.		129
b. 63		
c.		3

Table 2

X	Function X ÷ 2	Y
d.		34
e. 108		
f.		0

Table 3

X	Function 200 − X	Y
g.		147
h. 100		
i.		86

Lesson 120

Part 1

Table 1

X	Function X − 87	Y
a.		129
b. 87		
c.		87

Table 2

X	Function X + 87	Y
d.		87
e. 87		
f.		187

Table 3

X	Function X ÷ 9	Y
g.		81
h. 81		
i.		45

Part 2

a. $\frac{5}{1} \times \frac{19}{20} = \frac{95}{20}$

$\frac{5}{1}\frac{95}{20}$

b. $\frac{2}{3} \times \frac{28}{28} = \frac{56}{84}$

$\frac{2}{3}\frac{56}{84}$

c. $\frac{7}{4} \times \frac{13}{13} = \frac{91}{52}$

$\frac{7}{4}\frac{91}{52}$

d. $\frac{1}{29} \times \frac{5}{3} = \frac{5}{87}$

$\frac{1}{29}\frac{5}{87}$

e. $\frac{2}{15} \times \frac{6}{6} = \frac{12}{90}$

$\frac{2}{15}\frac{12}{90}$

f. $\frac{9}{5} \times \frac{37}{40} = \frac{333}{200}$

$\frac{9}{5}\frac{333}{200}$

Lesson 121

a. black dogs $\dfrac{3}{8}$ — of the _____ were _____.

b. [   ] — $\dfrac{9}{20}$ of the birds were pigeons.

c. [   ] — $\dfrac{5}{9}$ of the boys played soccer.

d. hot meals $\dfrac{2}{7}$ — of the _____ were _____.

e. [   ] — $\dfrac{1}{4}$ of the cars were SUVs.

Part 2

a. $\dfrac{46}{10}$ $\dfrac{14}{10}$ →

b. $\dfrac{8}{100}$ → $\dfrac{100}{100}$

c. _____ $\dfrac{5}{5}$ → $\dfrac{12}{5}$

_____

_____

_____

Part 3

a. $\dfrac{8}{10} - \dfrac{23}{100} =$

b. $2 + \dfrac{49}{100} + \dfrac{6}{10} =$

c. $\dfrac{2}{100} + \dfrac{4}{10} =$

Connecting Math Concepts

Lesson 122

Part 1

$$2 \times 2 \times 2 \times 5 = 40$$

a. $2 \times (2 \times 2 \times 5) = 40$ c. $5 \times (2 \times 2 \times 2) = 40$

b. $2 \times 2 \times (2 \times 5) = 40$ d. $2 \times 2 \times 2 \times (5) = 40$

Part 2

a. $\dfrac{15}{6} \longrightarrow \dfrac{31}{6}$ _____   c. $\longrightarrow \dfrac{9}{10} \longrightarrow \dfrac{64}{10}$ _____

b. $\dfrac{27}{8} \dfrac{34}{8} \longrightarrow$ _____   d. $\dfrac{19}{47} \longrightarrow \dfrac{47}{47}$ _____

Lesson 123

Part 1

$$2 \times 2 \times 3 \times 7 = 84$$

a. $2 \times ($ _____ $) = 84$ d. $2 \times 3 \times ($ _____ $) = 84$

b. $3 \times ($ _____ $) = 84$ e. $7 \times ($ _____ $) = 84$

c. $2 \times 2 \times ($ _____ $) = 84$ f. $2 \times 2 \times 3 \times ($ __ $) = 84$

Part 2

a. $\dfrac{27}{10} - \dfrac{5}{100} =$ b. $\dfrac{2301}{100} + 9 + \dfrac{8}{10} =$

c. $\dfrac{654}{10} - 17 =$

Lesson 124

$$3 \times 5 \times 7 =$$

a. $3 \times ($ $) =$

c. $7 \times ($ $) =$

b. $5 \times ($ $) =$

d. $3 \times 5 \times ($ $) =$

Part 2

$$2 \times 3 \times 3 \times 5 =$$

a. $2 \times ($ $) =$

d. $2 \times 3 \times ($ $) =$

b. $3 \times ($ $) =$

e. $3 \times 3 \times ($ $) =$

c. $5 \times ($ $) =$

f. $2 \times 5 \times ($ $) =$

Lesson 125

a. $\frac{46}{3}$ 13   1500% $\frac{26}{2}$

_____
_____
_____

Write the values from largest to smallest.

a. _____

_____

b. 9.1 $\frac{41}{4}$ 91% $\frac{100}{10}$

_____
_____
_____

Write the values from smallest to largest.

b. _____

_____

Part 2

$$2 \times 2 \times 5 \times 5 =$$

a. $2 \times ($ $) =$

c. $5 \times ($ $) =$

b. $2 \times 2 \times ($ $) =$

d. $2 \times 5 \times ($ $) =$

Part 3

$$2 \times 2 \times 2 \times 3 \times 3 =$$

a. $2 \times ($ $) =$

d. $2 \times 3 \times ($ $) =$

b. $3 \times ($ $) =$

e. $2 \times 2 \times 2 \times ($ $) =$

c. $2 \times 2 \times ($ $) =$

Lesson

Part 1

	feet	inches
ruler		12
yardstick	3	
rope	24	
chain		120

_____
_____
_____
_____
_____

Part 2

a. $\dfrac{740}{100}$ 7.4   743% $\dfrac{63}{9}$

_____
_____

Write the values from largest to smallest.

a. _____
_____

b. $\dfrac{33}{3}$ 12.1 $\dfrac{96}{9}$ 1121%

_____
_____

Write the values from smallest to largest.

b. _____
_____

Connecting Math Concepts

Lesson 127

Part 1

	weeks	days
spring break	2	
February		28
summer vacation	11	
school year		287

Lesson 128

Part 1

	gallons	quarts
milk jug	1	4
glass	$\frac{1}{8}$	
can		24
barrel	$31\frac{1}{2}$	
bin		93

Part 2

a. $8 \times 5 - \underline{\quad} = 40 - 10$

b. $24 - 63 \div 7 = \underline{\quad} - 9$

c. $\frac{16}{16} \times (15 + .8 + .03) = 15.83 \times \underline{\quad}$

d. $100 \div 25 + \underline{\quad} = 4 + 300 + 70 + 8$

Lesson

a. $91 - 10 + 15 = 9 \times \underline{} + 15$

b. $200 + \underline{} + 6 = 356 - 100$

c. $\dfrac{3}{8} = \dfrac{1}{8} + \dfrac{1}{8} + \underline{}$

d. $4\dfrac{2}{5} = 1 + \underline{} + \underline{} + \underline{} + \dfrac{1}{5} + \underline{}$

e. $\dfrac{5}{2} =$

f. $3\dfrac{4}{7} =$

Lesson

Part 1

a. $2\dfrac{5}{9} = 1 + \underline{} + \dfrac{1}{9} + \dfrac{1}{9} + \dfrac{1}{9} + \dfrac{1}{9} + \underline{}$

b. $\dfrac{4}{7} =$

c. $55 \div 5 - 10 = \underline{} - 10$

d. $3\dfrac{2}{3} =$

e. $500 + 70 + 4 - \underline{} = 474$

Part 2

$$3\overline{\smash)7{,}522}$$
2 5 0 7 $\tfrac{1}{3}$

3 ⟌ 7₁5 2 2 $\tfrac{1}{3}$

6   2 1

a. $7 \div 3 = \underline{}$ with a remainder of $\underline{}$.

b. $15 \div 3 = \underline{}$ with a remainder of $\underline{}$.

c. $2 \div 3 = \underline{}$ with a remainder of $\underline{}$.

d. $22 \div 3 = \underline{}$ with a remainder of $\underline{}$.

Lesson 130

Part 3

Weight of Packages

$\frac{5}{4}$ pounds $\frac{3}{4}$ pound

$\frac{3}{4}$ pound $\frac{2}{4}$ pound

$\frac{6}{4}$ pounds $\frac{3}{4}$ pound

$\frac{1}{4}$ pound $\frac{6}{4}$ pounds

$\frac{1}{4}$ $\frac{2}{4}$ $\frac{3}{4}$ $\frac{4}{4}$ $\frac{5}{4}$ $\frac{6}{4}$ $\frac{7}{4}$ pounds

Weight of Packages Line Plot

Questions

a. What is the weight of the heaviest package? _____

b. What is the weight of the lightest package? _____

c. What is the difference between the heaviest and the lightest package?

_____

d. What's the largest number of packages with the same weight?

_____

e. How heavy are the packages that have the largest number with the same weight? _____

f. What's the second-largest number of packages with the same weight?

_____

g. How heavy are the packages that have the second-largest number with the same weight? _____

h. What is the difference in weight between the packages having the largest number and the second-largest number with the same weight?

_____

Level E Correlation to Grade 4
Common Core State Standards for Mathematics

Operations and Algebraic Thinking (4.OA)

Use the four operations with whole numbers to solve problems.

1. Interpret a multiplication equation as a comparison, e.g., interpret $35 = 5 \times 7$ as a statement that 35 is 5 times as many as 7 and 7 times as many as 5. Represent verbal statements of multiplicative comparisons as multiplication equations.

Lessons	TB: 46–56, 58, 59, 61, 65, 66, 68, 69, 71, 72, 74, 78, 83, 87, 116

Operations and Algebraic Thinking (4.OA)

Use the four operations with whole numbers to solve problems.

2. Multiply or divide to solve word problems involving multiplicative comparison, e.g., by using drawings and equations with a symbol for the unknown number to represent the problem, distinguishing multiplicative comparison from additive comparison.

Lessons	TB: 46–48, 51–54, 56, 58–95, 108–118, 121, 124–127

Operations and Algebraic Thinking (4.OA)

Use the four operations with whole numbers to solve problems.

3. Solve multistep word problems posed with whole numbers and having whole-number answers using the four operations, including problems in which remainders must be interpreted. Represent these problems using equations with a letter standing for the unknown quantity. Assess the reasonableness of answers using mental computation and estimation strategies including rounding.

Lessons	TB: 113, 127, 130

Operations and Algebraic Thinking (4.OA)

Gain familiarity with factors and multiples.

4. Find all factor pairs for a whole number in the range 1–100. Recognize that a whole number is a multiple of each of its factors. Determine whether a given whole number in the range 1–100 is a multiple of a given one-digit number. Determine whether a given whole number in the range 1–100 is prime or composite.

Lessons	WB: 33, 36, 39, 40, 101–103, 106–110, 113, 114, 118, 119, 122–125 TB: 37–40, 100, 104, 111–130

Grade 4 Common Core State Standards Correlation

Connecting Math Concepts

Operations and Algebraic Thinking (4.OA)

Generate and analyze patterns.

5. Generate a number or shape pattern that follows a given rule. Identify apparent features of the pattern that were not explicit in the rule itself. *For example, given the rule "Add 3" and the starting number 1, generate terms in the resulting sequence and observe that the terms appear to alternate between odd and even numbers. Explain informally why the numbers will continue to alternate in this way.*

Lessons	WB: 13, 113, 114 TB: 115 Student Practice Software: Block 3 Activity 4, Block 4 Activity 5, Block 5 Activity 2

Number and Operations in Base Ten (4.NBT)

Generalize place value understanding for multi-digit whole numbers.

1. Recognize that in a multi-digit whole number, a digit in one place represents ten times what it represents in the place to its right. *For example, recognize that 700 ÷ 70 = 10 by applying concepts of place value and division.*

Lessons	WB: 2–5, 31–46, 48–51, 53–55, 57, 59, 60–66, 68–72, 75–82, 84, 92 TB: 5–9, 11–21, 23–26, 29, 35, 37, 39, 42, 45, 86, 114

Number and Operations in Base Ten (4.NBT)

Generalize place value understanding for multi-digit whole numbers.

2. Read and write multi-digit whole numbers using base-ten numerals, number names, and expanded form. Compare two multi-digit numbers based on meanings of the digits in each place, using >, =, and < symbols to record the results of comparisons.

Lessons	WB: 2–5, 7, 26–28, 31 TB: 5, 6, 28–31, 66, 67, 114–121

Number and Operations in Base Ten (4.NBT)

Generalize place value understanding for multi-digit whole numbers.

3. Use place value understanding to round multi-digit whole numbers to any place.

Lessons	WB: 81 TB: 80, 82–100, 106, 110, 112, 116, 121, 128, 130 Student Practice Software: Block 5 Activity 4

Number and Operations in Base Ten (4.NBT)

Use place value understanding and properties of operations to perform multi-digit arithmetic.

4. Fluently add and subtract multi-digit whole numbers using the standard algorithm.

Lessons	WB: 5–7, 19–22, 25, 26, 31, 36–39, 41, 43, 44, 56, 57, 81, 88, 114–121 TB: 3–6, 8–24, 27–80, 82–129 Student Practice Software: Block 1 Activity 4, Block 2 Activity 1, Block 5 Activities 4 and 5

Number and Operations in Base Ten (4.NBT)

Use place value understanding and properties of operations to perform multi-digit arithmetic.

5. Multiply a whole number of up to four digits by a one-digit whole number, and multiply two two-digit numbers, using strategies based on place value and the properties of operations. Illustrate and explain the calculation by using equations, rectangular arrays, and/or area models.

Lessons	WB: 2–6, 8, 10, 12–14, 16, 17, 19, 21, 22–46, 48–55, 57–70, 71, 72, 74, 75, 119, 120, 126, 127 TB: 6, 7, 9, 12–21, 23, 25–77, 79, 82, 84–99, 101–130 Student Practice Software: Block 1 Activity 4, Block 2 Activity 3, Block 3 Activity 5, Block 4 Activity 6, Block 5 Activity 5

Number and Operations in Base Ten (4.NBT)

Use place value understanding and properties of operations to perform multi-digit arithmetic.

6. Find whole-number quotients and remainders with up to four-digit dividends and one-digit divisors, using strategies based on place value, the properties of operations, and/or the relationship between multiplication and division. Illustrate and explain the calculation by using equations, rectangular arrays, and/or area models.

Lessons	WB: 21–23, 25, 53, 64–66, 72, 73, 74, 75, 104–110, 114, 119, 125, 127, 128, 130 TB: 22–40, 42, 45–121, 123, 124–130 Student Practice Software: Block 5 Activities 5 and 6

Number and Operations—Fractions (4.NF)

Extend understanding of fraction equivalence and ordering.

1. Explain why a fraction a/b is equivalent to a fraction $(n \times a)/(n \times b)$ by using visual fraction models, with attention to how the number and size of the parts differ even though the two fractions themselves are the same size. Use this principle to recognize and generate equivalent fractions.

Lessons	WB: 18, 35, 70–72, 119–121 TB: 19–27, 29, 31, 33, 40, 42, 73–114, 116, 119, 120, 122–126, 128 Student Practice Software: Block 2 Activity 4, Block 4 Activity 3, Block 5 Activity 5

Number and Operations—Fractions (4.NF)

Extend understanding of fraction equivalence and ordering.

2. Compare two fractions with different numerators and different denominators, e.g., by creating common denominators or numerators, or by comparing to a benchmark fraction such as 1/2. Recognize that comparisons are valid only when the two fractions refer to the same whole. Record the results of comparisons with symbols >, =, or <, and justify the conclusions, e.g., by using a visual fraction model.

Lessons	WB: 95–98, 100, 105–107 TB: 99, 101–103, 108, 111, 115, 118, 124, 129 Student Practice Software: Block 2 Activity 6, Block 3 Activity 1, Block 4 Activity 4

Number and Operations—Fractions (4.NF)

Build fractions from unit fractions by applying and extending previous understandings of operations on whole numbers.

3. Understand a fraction *a/b* with *a* > 1 as a sum of fractions 1/*b*.
   a. Understand addition and subtraction of fractions as joining and separating parts referring to the same whole.
   b. Decompose a fraction into a sum of fractions with the same denominator in more than one way, recording each decomposition by an equation. Justify decompositions, e.g., by using a visual fraction model. *Examples: 3/8 = 1/8 + 1/8 + 1/8 ; 3/8 = 1/8 + 2/8 ; 2 1/8 = 1 + 1 + 1/8 = 8/8 + 8/8 + 1/8.*
   c. Add and subtract mixed numbers with like denominators, e.g., by replacing each mixed number with an equivalent fraction, and/or by using properties of operations and the relationship between addition and subtraction.
   d. Solve word problems involving addition and subtraction of fractions referring to the same whole and having like denominators, e.g., by using visual fraction models and equations to represent the problem.

Lessons	WB: 28, 29, 31–34, 43, 44, 49, 54, 55, 59–66, 73–76, 80, 89, 90, 121–123, 129, 130 TB: 23–26, 30, 36, 37, 38, 44, 45, 46, 49–60, 64–79, 81–89, 91–95, 97–117, 119, 122–125, 127–129 **Student Practice Software: Block 2 Activity 2, Block 3 Activity 3, Block 4 Activity 1**

Number and Operations—Fractions (4.NF)

Build fractions from unit fractions by applying and extending previous understandings of operations on whole numbers.

4. Apply and extend previous understandings of multiplication to multiply a fraction by a whole number.
   a. Understand a fraction *a/b* as a multiple of 1/*b*. *For example, use a visual fraction model to represent 5/4 as the product 5 × (1/4), recording the conclusion by the equation 5/4 = 5 × (1/4).*
   b. Understand a multiple of *a/b* as a multiple of 1/*b*, and use this understanding to multiply a fraction by a whole number. *For example, use a visual fraction model to express 3 × (2/5) as 6 × (1/5), recognizing this product as 6/5. (In general, n × (a/b) = (n × a)/b.)*
   c. Solve word problems involving multiplication of a fraction by a whole number, e.g., by using visual fraction models and equations to represent the problem. *For example, if each person at a party will eat 3/8 of a pound of roast beef, and there will be 5 people at the party, how many pounds of roast beef will be needed? Between what two whole numbers does your answer lie?*

Lessons	WB: 98, 99, 102, 128 TB: 68–76, 78, 79, 81, 85, 89, 95–97, 100, 102–111, 114, 116, 117, 119, 121, 124, 125, 127, 129

Number and Operations—Fractions (4.NF)

Understand decimal notation for fractions, and compare decimal fractions.

5. Express a fraction with denominator 10 as an equivalent fraction with denominator 100, and use this technique to add two fractions with respective denominators 10 and 100. *For example, express 3/10 as 30/100, and add 3/10 + 4/100 = 34/100.*

Lessons	WB: 121, 123 TB: 122, 124

Number and Operations—Fractions (4.NF)

Understand decimal notation for fractions, and compare decimal fractions.

6. Use decimal notation for fractions with denominators 10 or 100. *For example, rewrite 0.62 as 62/100; describe a length as 0.62 meters; locate 0.62 on a number line diagram.*

Lessons	WB: 45–48, 126
	TB: 43, 44, 46, 47, 49–63, 66, 68, 70, 72, 74, 76, 96, 104, 108, 109, 114, 116, 120, 123, 125, 127, 129

Number and Operations—Fractions (4.NF)

Understand decimal notation for fractions, and compare decimal fractions.

7. Compare two decimals to hundredths by reasoning about their size. Recognize that comparisons are valid only when the two decimals refer to the same whole. Record the results of comparisons with the symbols >, =, or <, and justify the conclusions, e.g., by using **the number line or another** visual model.

Lessons	WB: 110–113
	eLessons: 111–116
	Student Practice Software: Block 2 Activity 6, Block 3 Activity 1, Block 4 Activity 4

*Denotes California-only content.

Measurement and Data (4.MD)

Solve problems involving measurement and conversion of measurements from a larger unit to a smaller unit.

1. Know relative sizes of measurement units within one system of units including km, m, cm; kg, g; lb, oz.; l, ml; hr, min, sec. Within a single system of measurement, express measurements in a larger unit in terms of a smaller unit. Record measurement equivalents in a two-column table. *For example, know that 1 ft is 12 times as long as 1 in. Express the length of a 4 ft snake as 48 in. Generate a conversion table for feet and inches listing the number pairs (1, 12), (2, 24), (3, 36), ...*

Lessons	WB: 126–128
	TB: 56–60, 63–66, 74, 75, 77, 82, 84

Measurement and Data (4.MD)

Solve problems involving measurement and conversion of measurements from a larger unit to a smaller unit.

2. Use the four operations to solve word problems involving distances, intervals of time, liquid volumes, masses of objects, and money, including problems involving simple fractions or decimals, and problems that require expressing measurements given in a larger unit in terms of a smaller unit. Represent measurement quantities using diagrams such as number line diagrams that feature a measurement scale.

Lessons	TB: 12–22, 24, 25–54, 56–74, 76–80, 82–94, 105–108, 110–113, 115, 116, 118, 120–130

Measurement and Data (4.MD)

Solve problems involving measurement and conversion of measurements from a larger unit to a smaller unit.

3. Apply the area and perimeter formulas for rectangles in real world and mathematical problems. *For example, find the width of a rectangular room given the area of the flooring and the length, by viewing the area formula as a multiplication equation with an unknown factor.*

Lessons	WB: 7–10 TB: 5, 6, 9, 11–18, 20, 21–24, 27, 29, 31, 33, 34, 37, 39–41, 44, 48, 53, 55, 58, 60, 62, 71, 75, 82, 85, 88–102, 104, 106, 108, 110, 112, 114, 117, 121, 124–127, 129 Student Practice Software: Block 1 Activity 6

Measurement and Data (4.MD)

Represent and interpret data.

4. Make a line plot to display a data set of measurements in fractions of a unit (1/2, 1/4, 1/8)..Solve problems involving addition and subtraction of fractions by using information presented in line plots. *For example, from a line plot find and interpret the difference in length between the longest and shortest specimens in an insect collection.*

Lessons	WB: 130 TB: 129

Measurement and Data (4.MD)

Geometric measurement: understand concepts of angle and measure angles.

5. Recognize angles as geometric shapes that are formed wherever two rays share a common endpoint, and understand concepts of angle measurement:

a. An angle is measured with reference to a circle with its center at the common endpoint of the rays, by considering the fraction of the circular arc between the points where the two rays intersect the circle. An angle that turns through 1/360 of a circle is called a "one-degree angle," and can be used to measure angles.

b. An angle that turns through *n* one-degree angles is said to have an angle measure of *n* degrees.

Lessons	WB: 68, 78, 79 TB: 63–67, 69–84, 87, 90, 91, 93, 95, 97, 99, 103, 107–110, 112–119, 121–127

Measurement and Data (4.MD)

Geometric measurement: understand concepts of angle and measure angles.

6. Measure angles in whole-number degrees using a protractor. Sketch angles of a specified measure.

Lessons	TB: 121–127

Measurement and Data (4.MD)

Geometric measurement: understand concepts of angle and measure angles.

7. Recognize angle measure as additive. When an angle is decomposed into non-overlapping parts, the angle measure of the whole is the sum of the angle measures of the parts. Solve addition and subtraction problems to find unknown angles on a diagram in real world and mathematical problems, e.g., by using an equation with a symbol for the unknown angle measure.

Lessons	WB: 78, 79 TB: 69–84, 87, 90, 93, 97, 99, 103, 109, 112, 115, 123, 124, 126

Geometry (4.G)

Draw and identify lines and angles, and classify shapes by properties of their lines and angles.

1. Draw points, lines, line segments, rays, angles (right, acute, obtuse), and perpendicular and parallel lines. Identify these in two-dimensional figures.

Lessons	WB: 95–97, 100–103 TB: 110–120, 122, 125–127, 129, 130 **Student Practice Software: Block 5 Activity 1**

Geometry (4.G)

Draw and identify lines and angles, and classify shapes by properties of their lines and angles.

2. Classify two-dimensional figures based on the presence or absence of parallel or perpendicular lines, or the presence or absence of angles of a specified size. Recognize right triangles as a category, and identify right triangles.
   *(Two-dimensional shapes should include special triangles, e.g., equilateral, isosceles, scalene, and special quadrilaterals, e.g., rhombus, square, rectangle, parallelogram, trapezoid.)

Lessons	TB: 110–119, 122, 125–130 eLessons: 120–129 **Student Practice Software: Block 1 Activity 5**

*Denotes California-only content.

Geometry (4.G)

Draw and identify lines and angles, and classify shapes by properties of their lines and angles.

3. Recognize a line of symmetry for a two-dimensional figure as a line across the figure such that the figure can be folded along the line into matching parts. Identify line-symmetric figures and draw lines of symmetry.

	Student Practice Software: Block 3 Activity 2, Block 4 Activity 2, Block 5 Activity 3

Standards for Mathematical Practice

Connecting Math Concepts addresses all of the Standards for Mathematical Practice throughout the program. What follows are examples of how individual standards are addressed in this level.

1. Make sense of problems and persevere in solving them.

Addition and Multiplication Number Families (Lessons 1–130): Students learn to identify specific types of word problems (i.e., comparison, sequence, classification) and set up number families to solve the problems based on the specific problem types.

2. Reason abstractly and quantitatively.

Fractions (Lessons 1–130): Students' early work with fractions involves deriving fractions from pictures. They relate pictures of objects, such as a circle with two of three parts shaded, to the numbers for the numerator (parts shaded) and for the denominator (parts in the whole).

3. Construct viable arguments and critique the reasoning of others.

Prime Factors (Lessons 100–130): Students learn about prime and composite numbers and rules for divisibility. They can explain how they know a given number is divisible by another number. Students tell how they know 81 is divisible by 3 without doing any computation (Lesson 117).

4. Model with mathematics.

Addition and Multiplication Number Families (Lessons 1–130): Students learn to represent three related numbers in a number family. Later, they apply the number-family strategy to model and solve a variety of word problems involving whole numbers and fractions.

5. Use appropriate tools strategically.

Throughout the program (Lessons 1–130) students use pencils, workbooks, lined paper, and textbooks to complete their work. They use protractors to measure angles. Students also use the computer to access the Practice Software where they apply the skills they learn in the lessons.

6. Attend to precision.

Geometry (Lessons 12–34, 76–87, 106–118): When finding area, perimeter, and angle measurements, students attend to the units and respond verbally and in written answers with the correct unit. They also measure angles precisely to the degree using protractors. They include units in answers to word problems that involve specific units.

7. Look for and make use of structure.

Angles (Lessons 62–130): Students calculate measurements of missing angles using their knowledge of different types of angles, including right, supplementary, and vertical.

8. Look for and express regularity in repeated reasoning.

Coordinate System and Functions (Lessons 85–130): Starting in Lesson 94, students find missing Y values to complete a function table. Then they plot the points and see that they always result in a straight line. Students see that the function tells about all the points on the line.